The Surgeons Are Tied Up In The Operating Room

- A Literal Interpretation Of Surgery -

Peter C. Johnson, MD

The Surgeons Are Tied Up In The Operating Room

ISBN: 978-1-4357-1720-6

Printed in the United States of America

Disclaimer: The images in this book do not represent any particular persons or places. Comments or questions can be addressed to the author at pjohnson@scintellix.com

Additional copies of this book can be purchased at:
www.lulu.com
http:///www.lulu.com/content/2436310

Other books by Peter C. Johnson, MD

<u>At The Table</u>
http://www.lulu.com/content/2474191

The Surgeons Are Tied Up In The Operating Room is dedicated to the following:

My wife Karen, for living through multiple residencies with me.

The entire staff of the Case Western Reserve University Department of Surgery, all of its alumni and present trainees.

All of the healthcare providers in the CWRU system including University Hospitals of Cleveland, MetroHealth Medical Center, Rainbow Babies and Children's Hospital and the Veteran's Administration Hospital of Cleveland, Wade Park.

And finally to:

All of my patients and all surgeons, worldwide.

THIS BOOK
IS JUST WHAT
I'VE NEEDED...

Introduction

I presented the cartoons that make up The Surgeons Are Tied Up In The Operating Room at Surgery Grand Rounds at Case Western Reserve University on June 27, 1987. It was the last day of my seven year long General Surgery residency.

Normally a staid educational affair, Grand Rounds on that day was whimsical and fun—and full of the emotions one feels when departing from great colleagues—in some cases forever. The presentation itself was broken down into four reasonably cogent "case presentations." No such order is presented here because, frankly, I can't remember the sequence used. What I do remember is that the younger persons in the audience appeared to appreciate the talk while the more senior persons present ensured that I was whisked off to my next station in life as quickly as possible.

Those were the days before cell phones and personal digital assistants when residents with organized minds clearly differentiated themselves from the others. I was not one of the former, so I maintained hardcopy pocket-sized notebooks that held all manner of information related to the patients on service at the time. Surgical training being what it is, there were moments of chaos, hard work and boredom, randomly intertwined. It was during the periods of boredom that I began to imagine what our patients must think when we say things like "Blunt Chest Trauma," "Spilling Sugar in the Urine," "M and M Rounds" and the like. I created sketches in those books to illustrate ways in which they might interpret our jargon. Those sketches were the basis for the formal inked cartoons you see here. The notebooks, thankfully, survive.

Though I went on to a residency and ten years of practice in reconstructive Plastic Surgery, I no longer practice surgery. Nonetheless, in looking back at these cartoons, I immediately reconnect with their moments of creation—recollecting the joys, frustrations, fatigue and responsibility associated with caring for very sick people so intensely and for so long.

Becoming a surgeon is one of life's powerful experiences. I know, looking back, that retaining one's sense of humor in the process is important. Sketching humorous incidents helped me to do that.

With this in mind, I would like to share these cartoons with you. The book is structured so that each cartoon is depicted as a literal translation of the terminology that inspired it. Situated below each cartoon is an actual medically-correct description of what the terminology means. Sometimes, these are editorialized for emphasis.

While this is therefore a pictorial Encyclopedia of Surgery of sorts, I would not recommend it as a substitute for any venerable standard textbook of Surgery.

I hope you enjoy it.

Peter C. Johnson, MD
Raleigh, NC 2008

The Academic Surgeon

The academic surgeon is an unusual person. He or she is a rare combination of scholarship, aggression and direct action - as shown here in a reflective moment.
A variant more inclined to lab investigation may be found in a white coat surrounded by 100 dead rats
- and no hypothesis.

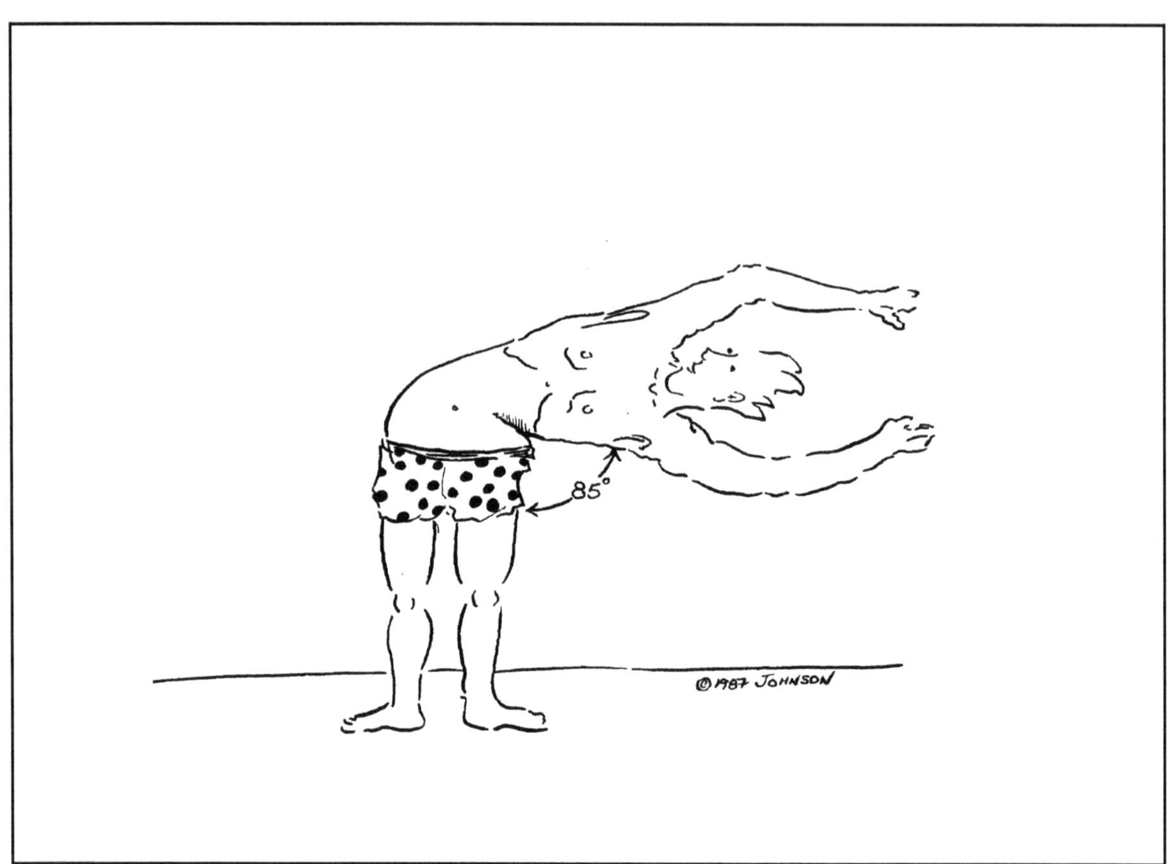

The Acute Abdomen

The acute abdomen is a classical surgical emergency characterized by extreme tenderness, often an elevated white blood cell count and low grade fever. Appendicitis is a classic cause. If a patient can assume the posture shown, suspect the absence of an acute abdomen.

Ambulatory Surgery

One of the great misnomers in medicine, ambulatory surgery is a title that does not apply to the surgery itself but to the fact that the patient somehow survives surgery and can still walk away the same day.

Anesthesia By Mask

Usually performed for short operations, anesthesia by mask requires the anesthesiologist to hold a rubber mask tightly to the face, hold up the jaw and squeeze the ventilator bag regularly.

This is generally not their first choice.

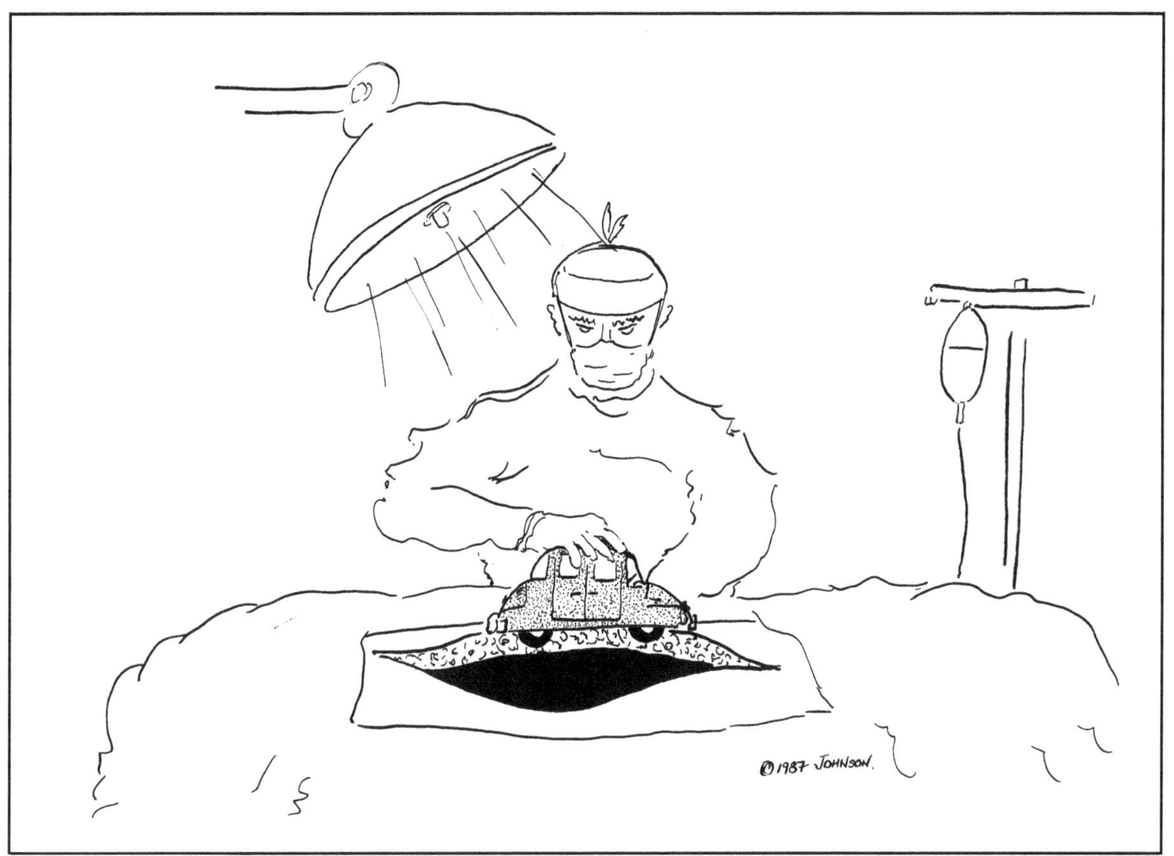

Autotransplantation

Autotransplantation is treatment with one's own tissues. Skin grafting, transfusion of one's own saved blood or stem cells are examples.

A Volkswagen does not qualify.

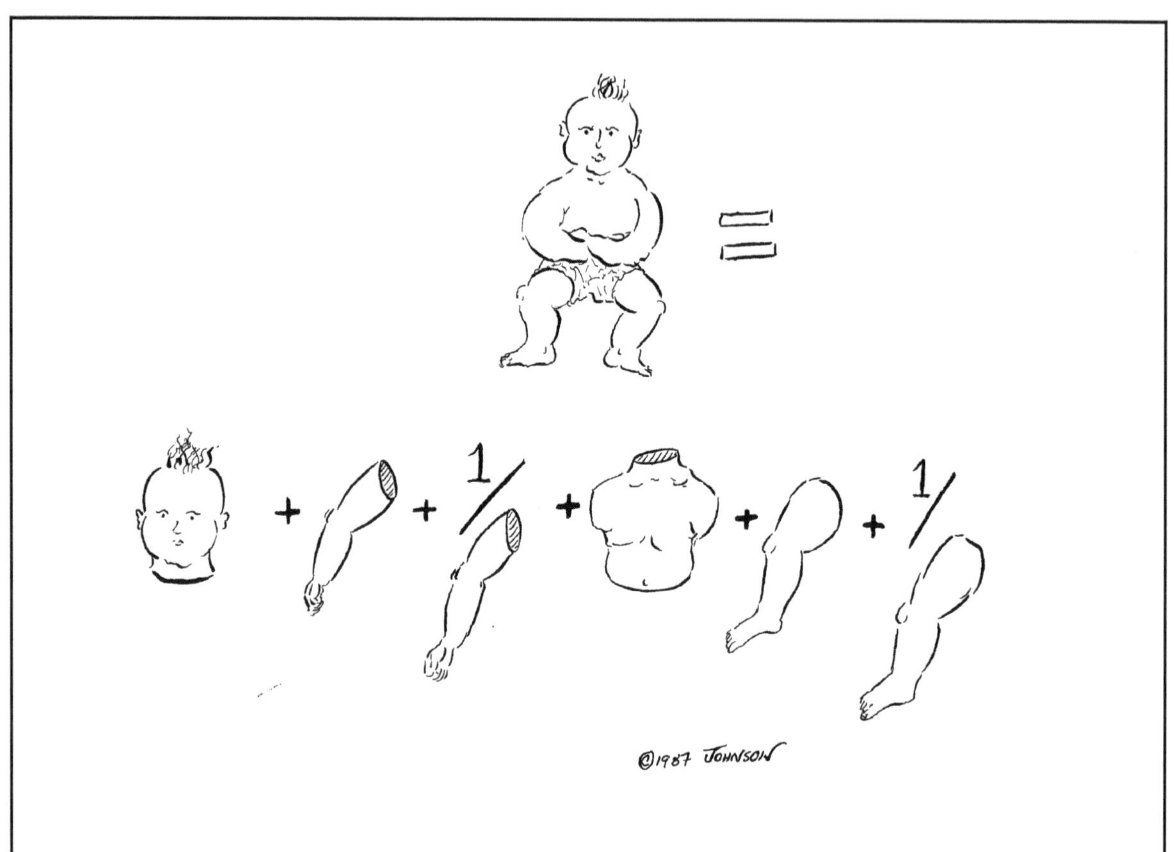

Baby Formula

Highly nutritious feeding supplement for babies whose use irritates ardent supporters of breast feeding to no end.

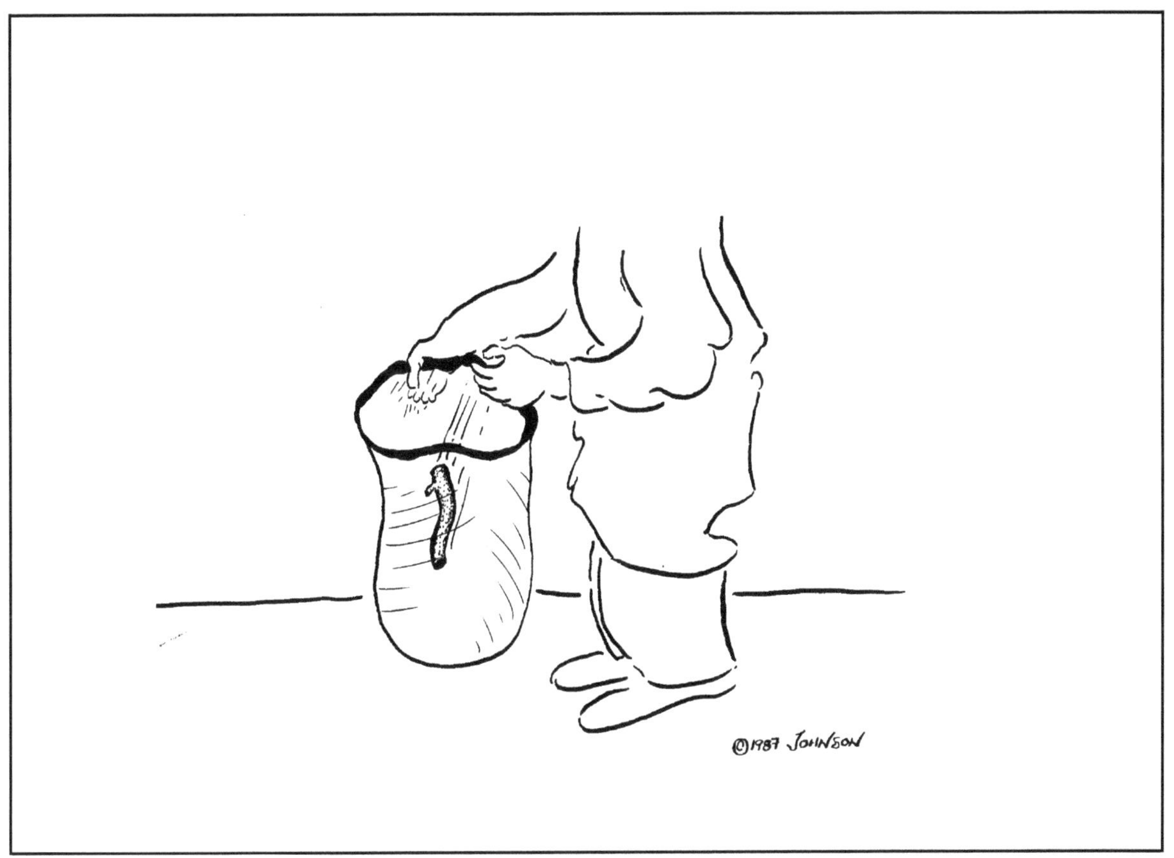

Bagging The Common Duct

This is terminology from the surgeons' locker room. "Charlie bagged the common duct again" means that Charlie will be at Morbidity and Mortality conference next month to state why he cut across a sacred biliary structure inadvertently.
A major no-no.

Bloodmobile

Without the bloodmobile, surgeons could not operate. Years of searching for this as a vanity license plate on a surgeon's car have gone for naught.

Perhaps BLDMBL?

Blunt Chest Trauma

One of the most common presentations of trauma, blunt chest trauma is most feared as a cause of cryptic aortic transection and severe pulmonary contusions (bruising).

The Breast Pump

A recent mother who is lactating is able to relieve, rather than increase breast pressure using a breast pump—and thereby store milk for her child.

Calling A Code

Calling a code over the paging system alerts an emergency medical response team that a cardiac arrest or other emergency has occurred. Generally, the location is provided. It triggers one of the few accepted reasons for running in a hospital.
The other is to eat between operations.

The Cards Consult

The cardiologist is often consulted to assess the patient's heart health before complex operations are undertaken. The result is a magical Tarot-like prediction of future cardiac events that generally delay surgery.

The Charge Nurse

The acting 'chief operating officer' of the ward, the charge nurse is the go-to person for problem solving. An early component of the surgeon's education is to learn to never cross the charge nurse.

The Chest Is Clear

When a chest x-ray is ordered because the surgeon is worried about pneumonia and the radiologist declares "The chest is clear," it is cause for celebration.

Child Bucking The Ventilator

Nothing is more poignant than a sick child. Often, their fears make them resist our attempts at treatment. When on ventilators, they often gamely fight—and win.

Child Passing Formed Stools

A welcome moment for the pediatric surgeon watching intently for postoperative success, formed stools indicate an intact and functioning gastrointestinal tract.

Cracking The Chest

Though gruesome, sometimes in the face of a cardiac arrest or massive internal bleeding, the chest must be entered without anesthesia to apply cardiac massage or to apply direct pressure to bleeding vessels.
The patient is often left with a quizzical look as this process unfolds.

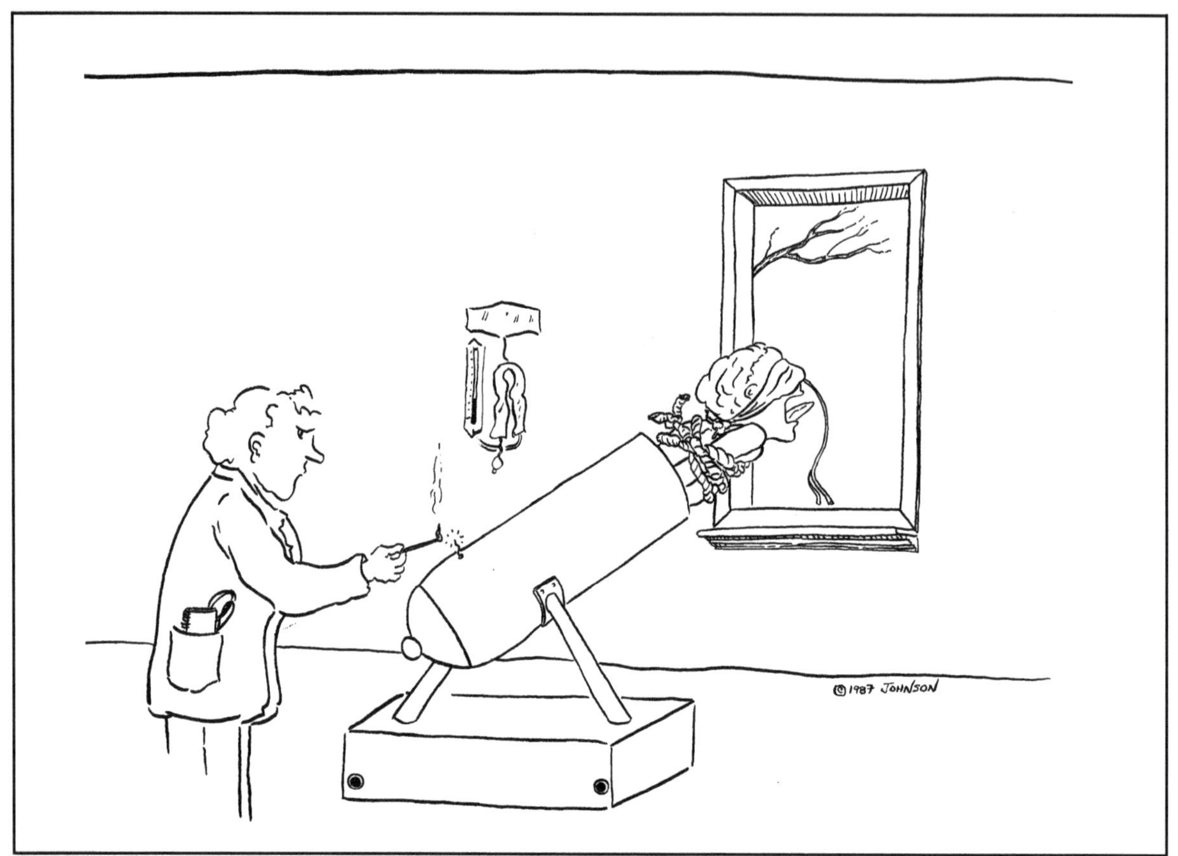

Discharging The Patient

Actually quite a paperwork ordeal, the delights of sending a patient on his or her way is often left to the intern.

Entering The Patient Into A Trial

"Clinical trials of therapies are essential to further our ability to improve the care of all patients," the MD is overheard saying to this scantily clad, trusting fellow.

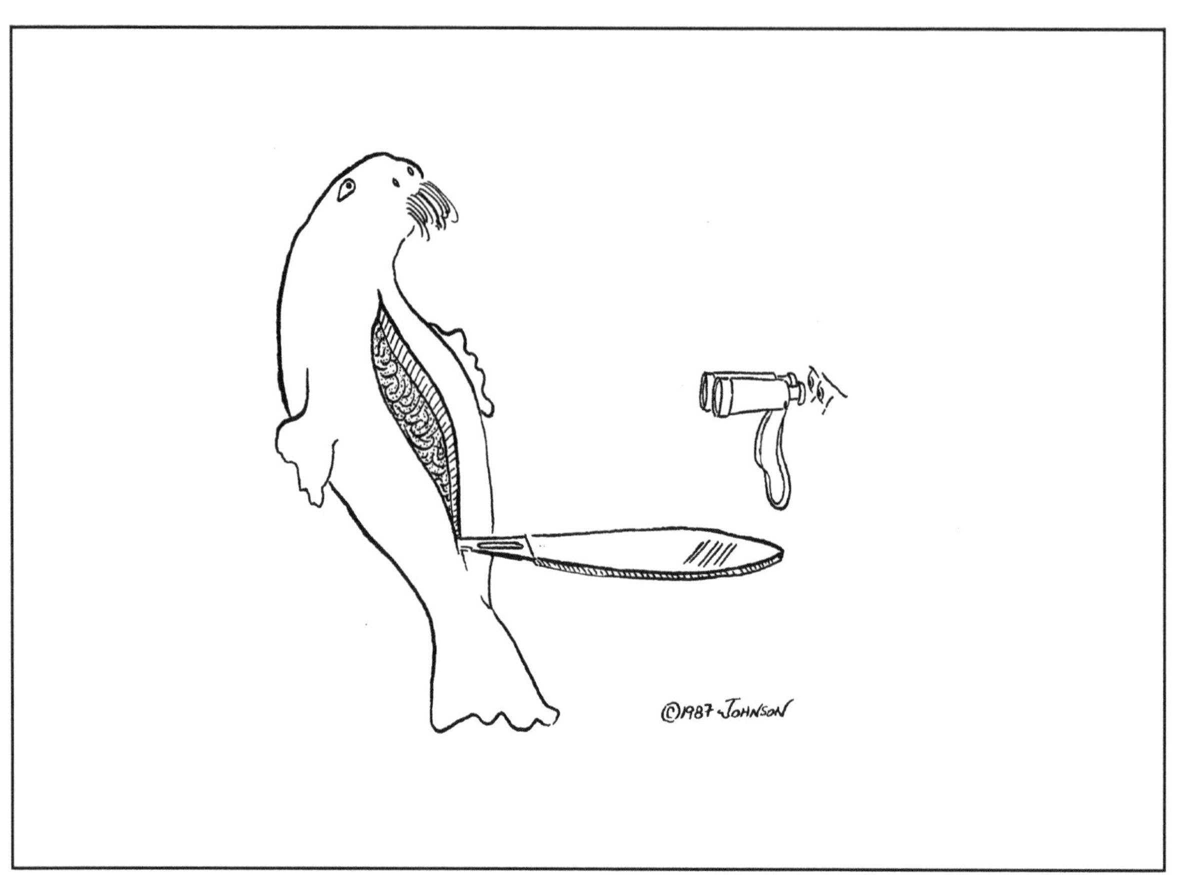

Exploratory Celiotomy

The abdomen is technically a coelom (an internal body cavity). A celiotomy is therefore, simply a peek inside, generally done to determine the presence of appendicitis, another inflammatory disorder or a tumor. The seal shown here is resigned to this.

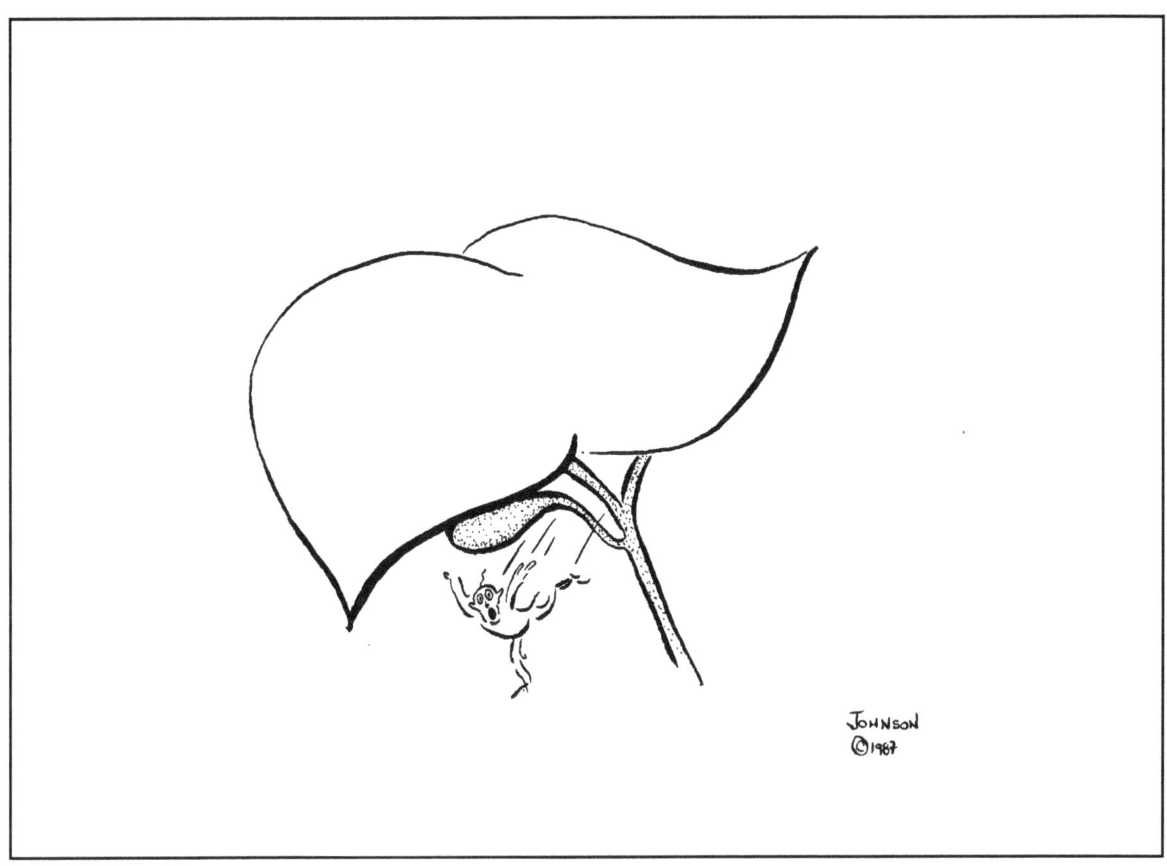

Falling Out Of The Biliary Tree

A patient may be jaundiced (green) for many reasons but one of the most common is when stones being formed in the branches of the biliary system fall down and block the junction of the common bile duct and the small bowel.

A Femoral Head, Collapsed And Subluxed

Classic orthopedic terminology, collapse and subluxation of the femoral head simply means that during a fracture of the hip, the head of the thigh bone at its junction with the pelvis has been compressed and shifted from its normal position.

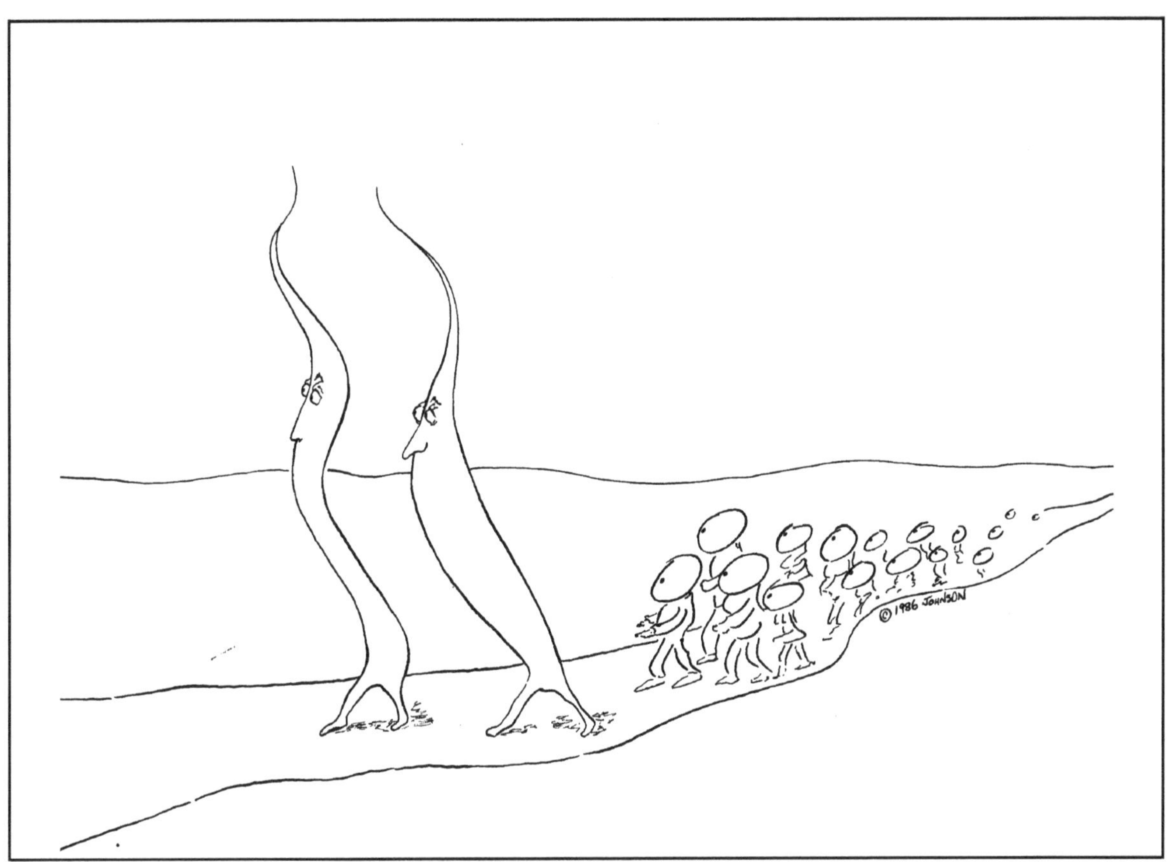

Filiforms And Followers

When one cannot urinate due to stricture (tightening) of the urethra, fine, linked polymeric dilators, graduating steadily in size are inserted through the obstruction to dilate it.

Relief is rapid and often requires a change of scrubs.

Following The Patient

Return visits are essential to track a patient's progress after surgery.

This surgeon is unusually watchful.

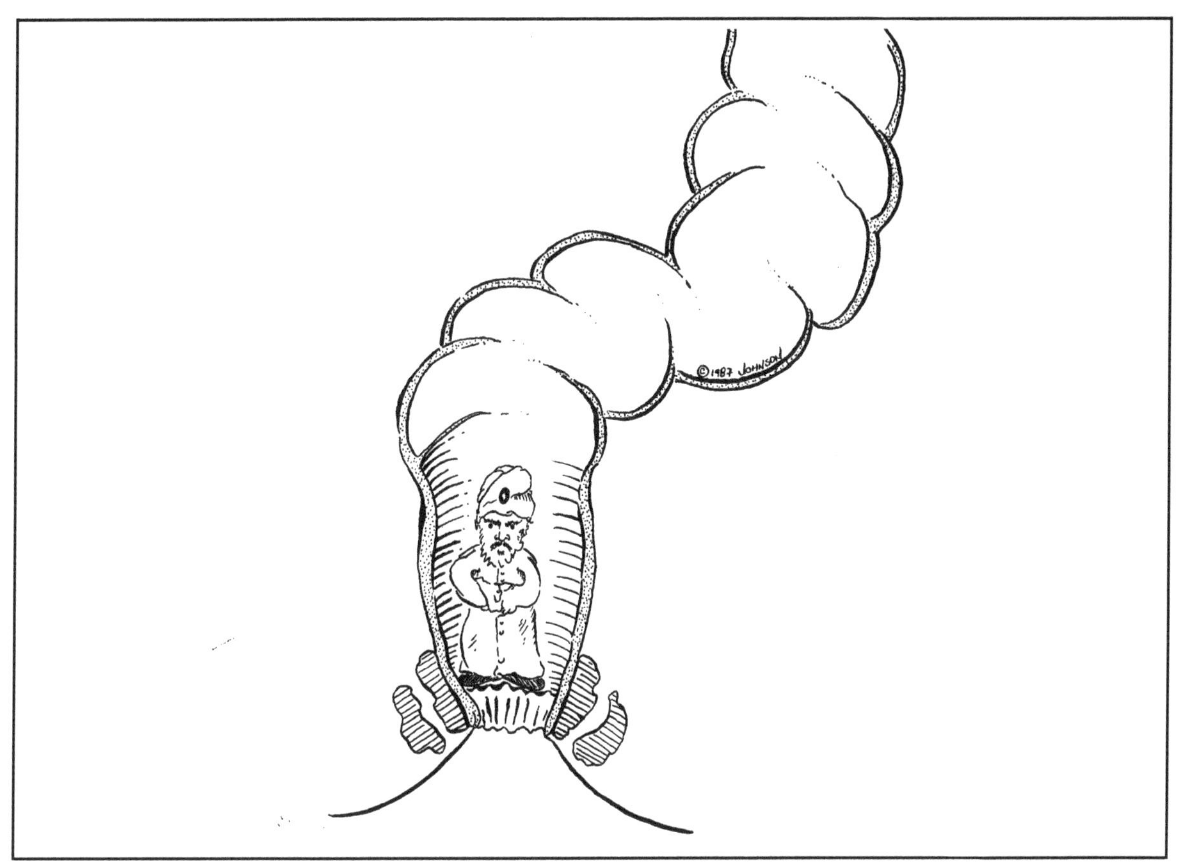

Foreign Body In The Rectum

Every MD has a story about being called to the emergency room to remove various and sundry objects from an embarrassed patient's rectum. This is best done silently, swiftly and without judgment.

Frank Pus

It is said that a surgeon draining an abscess is in his or her finest hour. Rarely can one do so much for another person in so little time. Often the first operation that an unsuspecting intern performs, the experience of releasing pus under pressure can be memorable.

Gastroscopy

A type of upper gastrointestinal endoscopy, gastroscopy evaluates the integrity of the stomach.

Going Over A Film With The Radiologist

Usually, x-rays are vividly illustrative of conditions and a quick look informs all. Sometimes, though, the subtle eye of the radiologist is needed to uncover hidden issues.
Surgeons do not like to ask directions so these moments are often tense.

Grand Rounds

All academic medical units hold a weekly educational conference known as 'Grand Rounds' during which the learned opine in circular fashion until the coffee and cookies run out.

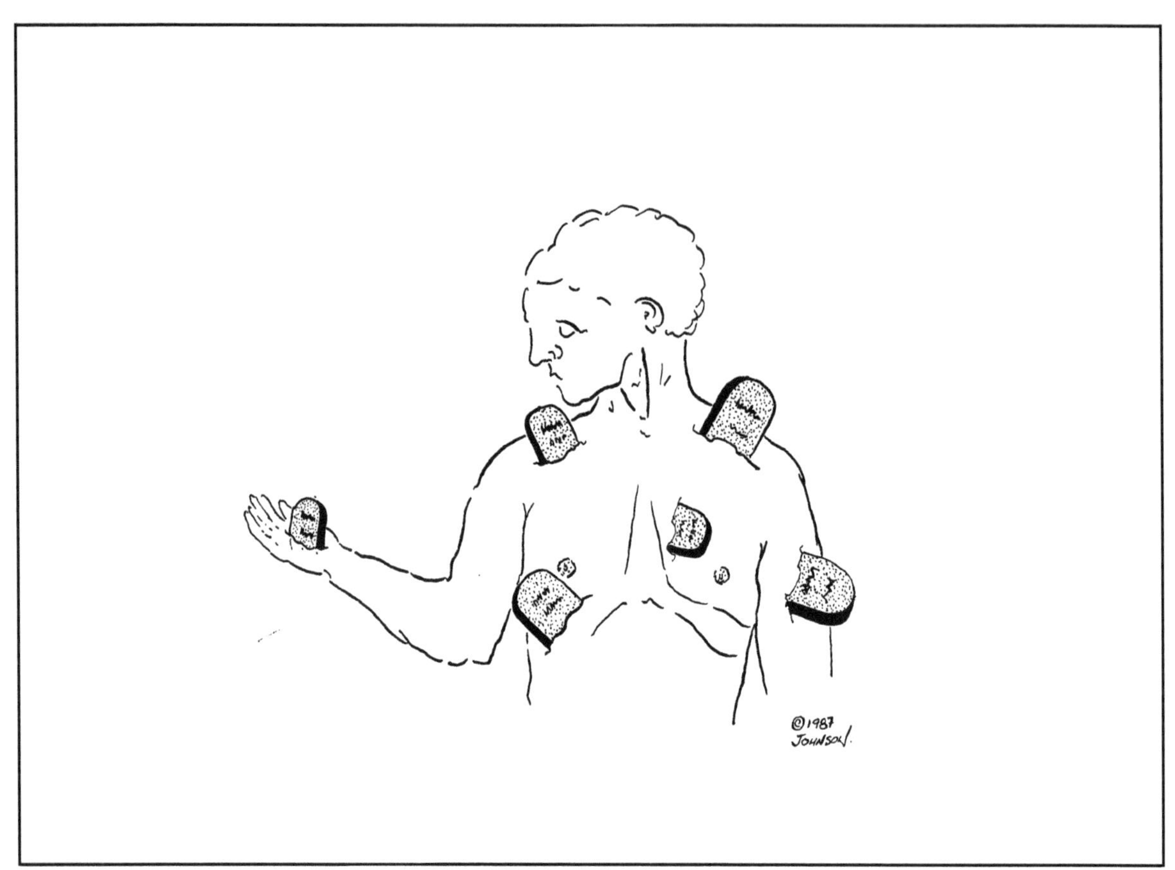

Grave's Disease

Grave's disease is autoimmune-mediated overactivity of the thyroid gland, leading to an increased metabolism, enlarged thyroid and bulging eyes. The name has an unfortunate 'terminal' ring to it.

The Heart Attack

An acute myocardial infarction is a heart attack. It is the sudden obstruction of a blood vessel to the heart and can be instantly fatal.

The High Riding Prostate

During severe pelvic trauma, the urethra can be transected. This is diagnosed on rectal exam by the presence of a 'floating' or 'high riding' prostate.

Images of the wild west cannot be avoided.

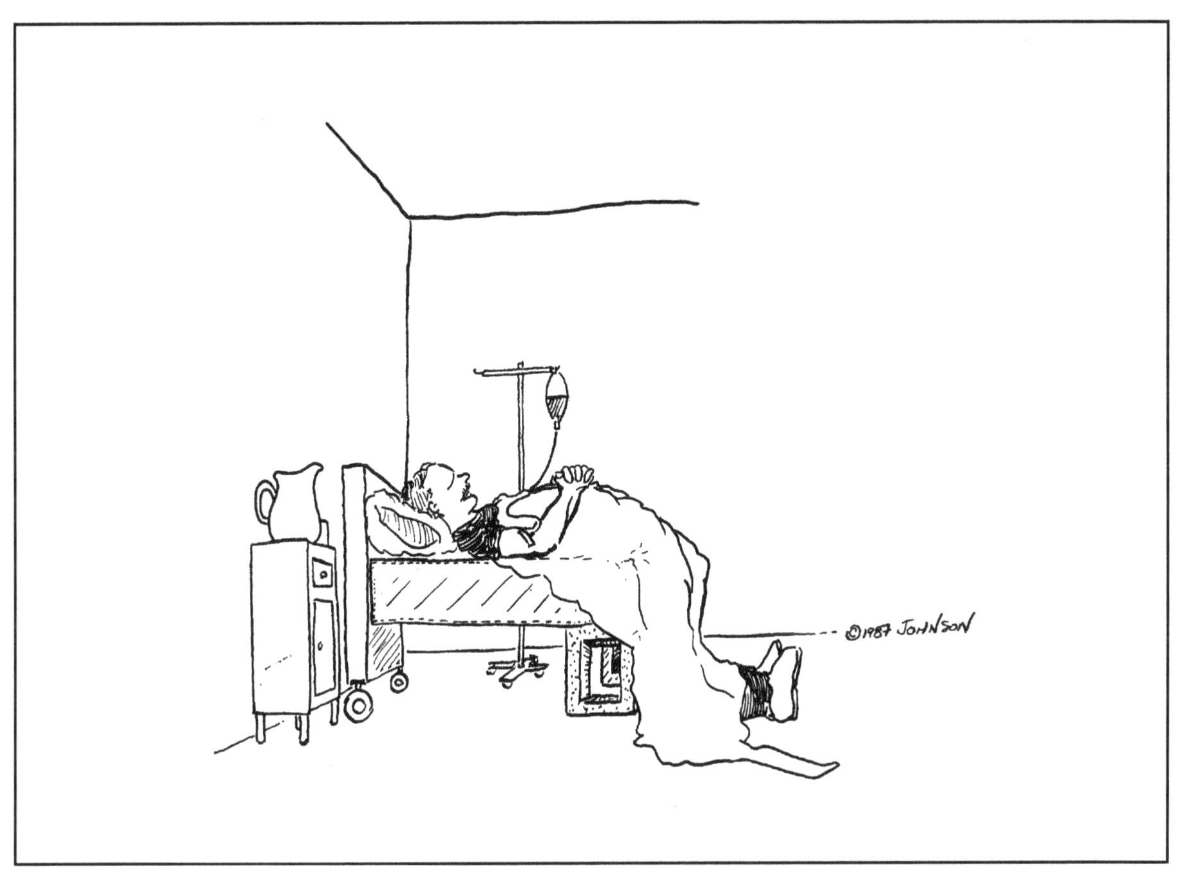

The Hospital Had A Severe Bed Shortage

Hospitals often fill to the extent that no additional patients can be admitted, a situation that is known as a 'bed shortage.'

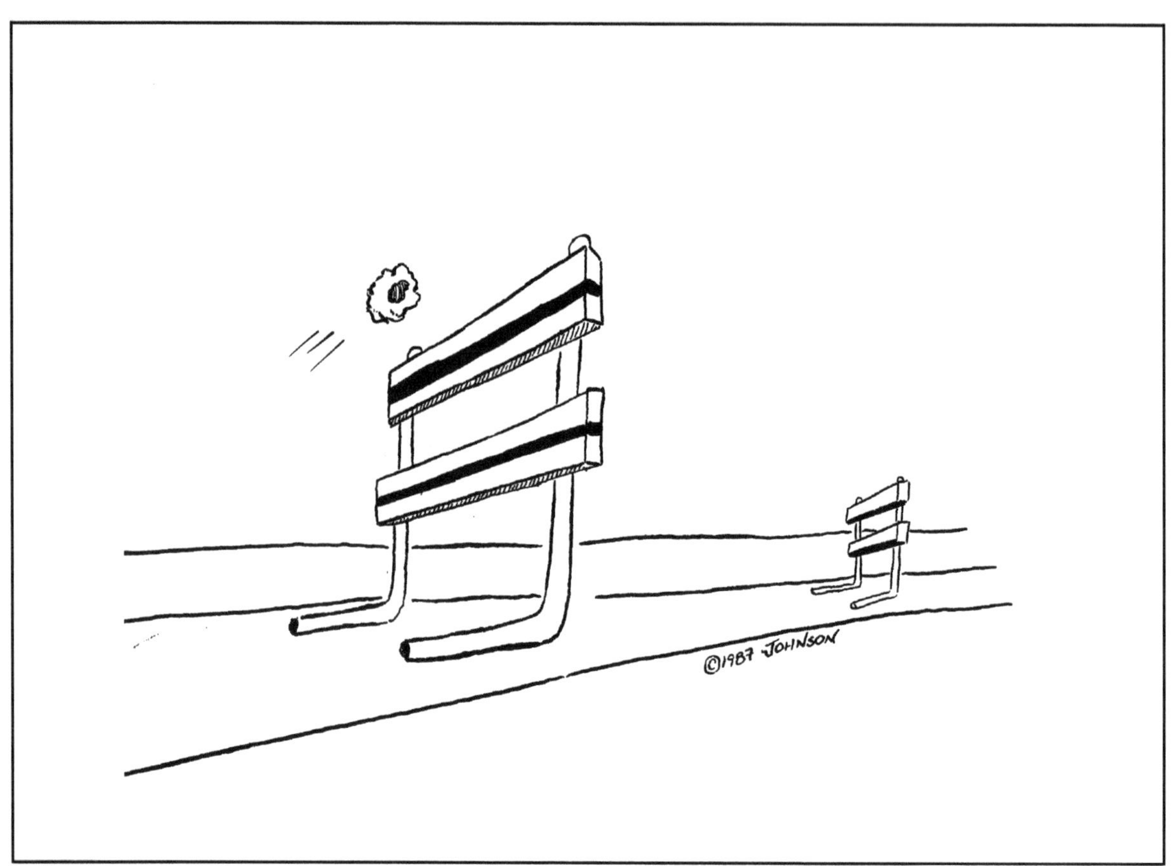

The Hurthle Cell

An enlarged epithelial cell of the thyroid, the Hurthle cell (pronounced 'hurdle') is associated with a form of thyroiditis known as 'Hashimoto's Thyroiditis' and also with some forms of thyroid cancer.

Hyperal

A patient who is unable to eat often needs to be fed by vein. When a patient's complete caloric requirement is administered by vein, this is termed 'hyperalimentation' or simply 'hyperal.' This term is also used to describe excitable surgeons named 'Al.'

The I.D. Consult

An Infectious Disease consultation is often requested whenever a serious infection is being treated with antibiotics. It is generally an intellectual affair.

Incentive Spirometry

A spirometer measures the ventilation of the lungs. After surgery or during prolonged bedrest, the lungs collapse slightly. A Respiratory Therapist is then called to administer breathing treatments to reinflate the lungs—usually with extreme prejudice.

The K+ Rider

Patients on diuretics often arrive at the hospital in a potassium (K+)-depleted state. Since an overdose of potassium can be fatal, it is often administered through an IV line in a small, controlled volume known by this name. There is no relationship to the high riding prostate, shown previously.

Lithotripsy

In Greek, 'lithos' means 'stone' and 'tripsis' means 'a rubbing.' Lithotripsy is the application of sound waves to break up kidney stones so that they may be passed in the urine.

M&M Rounds

Morbidity and Mortality Rounds are confidential gatherings of surgeons to discuss complications of surgery. It is a vital part of the 'Forgive and Remember' aspect of professional education. Criticism occurs. Pain is felt. Learning ensues.

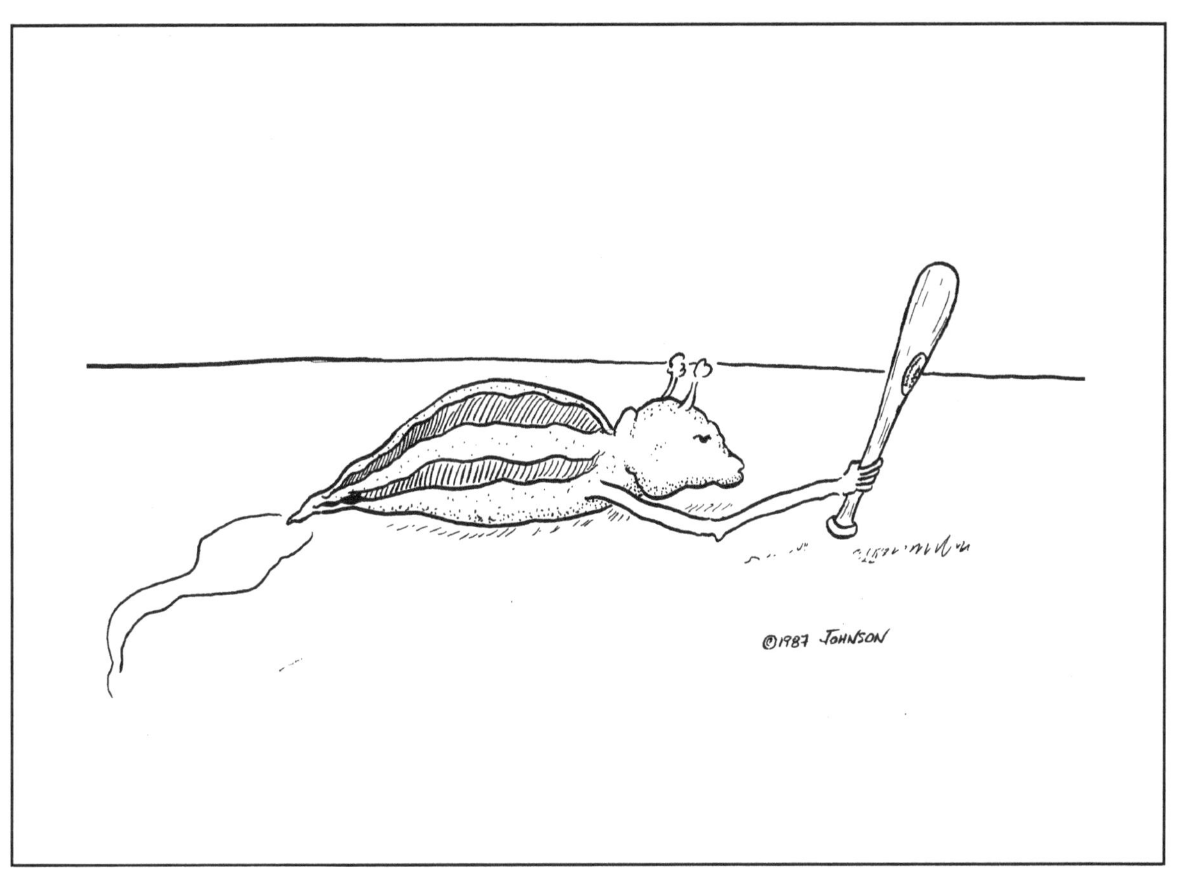

The Major League Slug

An otherwise healthy patient who remains bedbound postoperatively for longer than the routine period is often termed a 'Slug.' A 'Major League Slug' is one who turns such behavior into a profession, thereby prolonging hospitalization excessively.

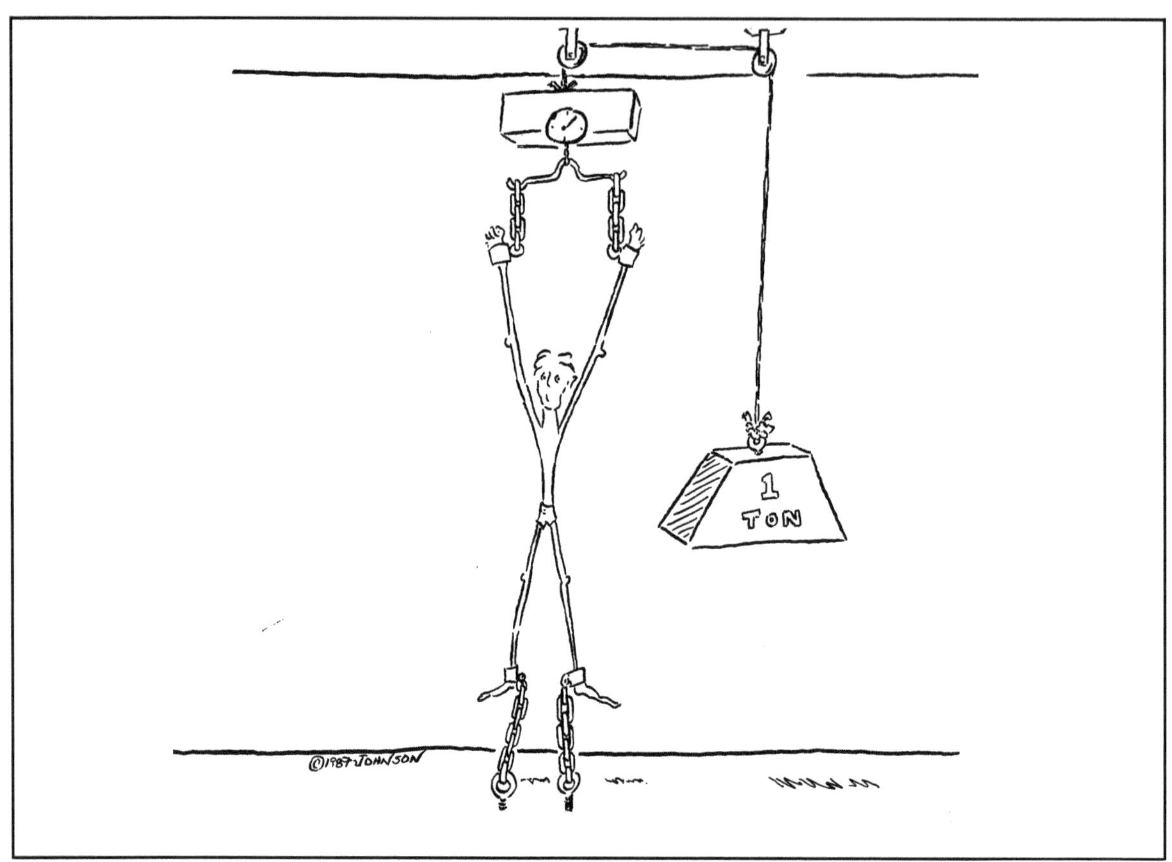

Measuring Patient Compliance

When patients are given prescriptions or activity regimens by their physicians, they often fail to take their medicines or follow other instructions. At times, they are quizzed heartily by the medical team to discern how dutiful they have been. This process can be taken to extremes.

Moon Facies

One of the complications of prolonged corticosteroid use is the development of an enlarged, rounded face, a condition known as 'Moon Facies.'

Athletes on steroids do not develop this because they use a different type of steroid.

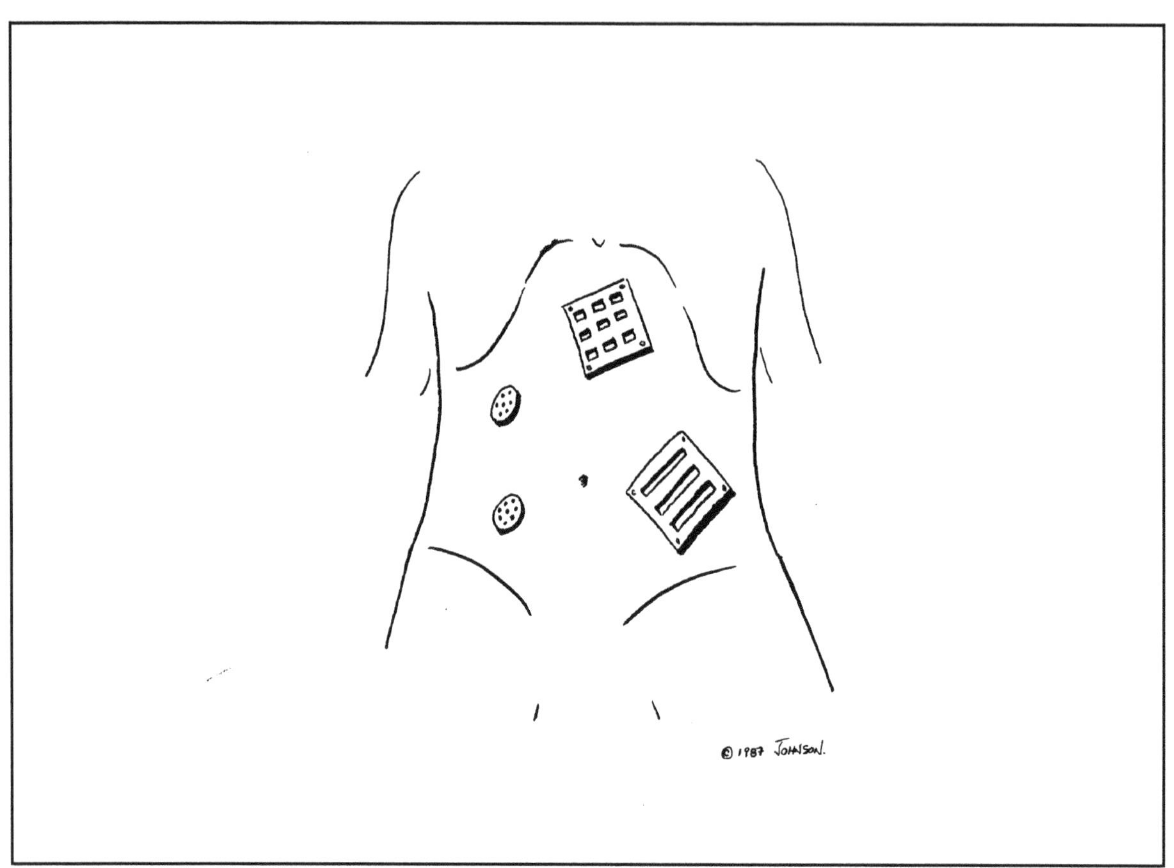

Multiple Drains Were Placed

The surgeon who is worried about postoperative fluid collection and infection in the abdomen will often place drains through the abdominal wall to allow fluids to flow out. The statement ‘multiple drains were placed’ is therefore often found in operative notes.

The Nasal Trumpet

A common device used when anesthesia is being given by mask, a nasal trumpet is a soft silicone tube with a flared front end that is lubricated and placed through a nostril to allow a ready access to the throat so that secretions can be suctioned away.

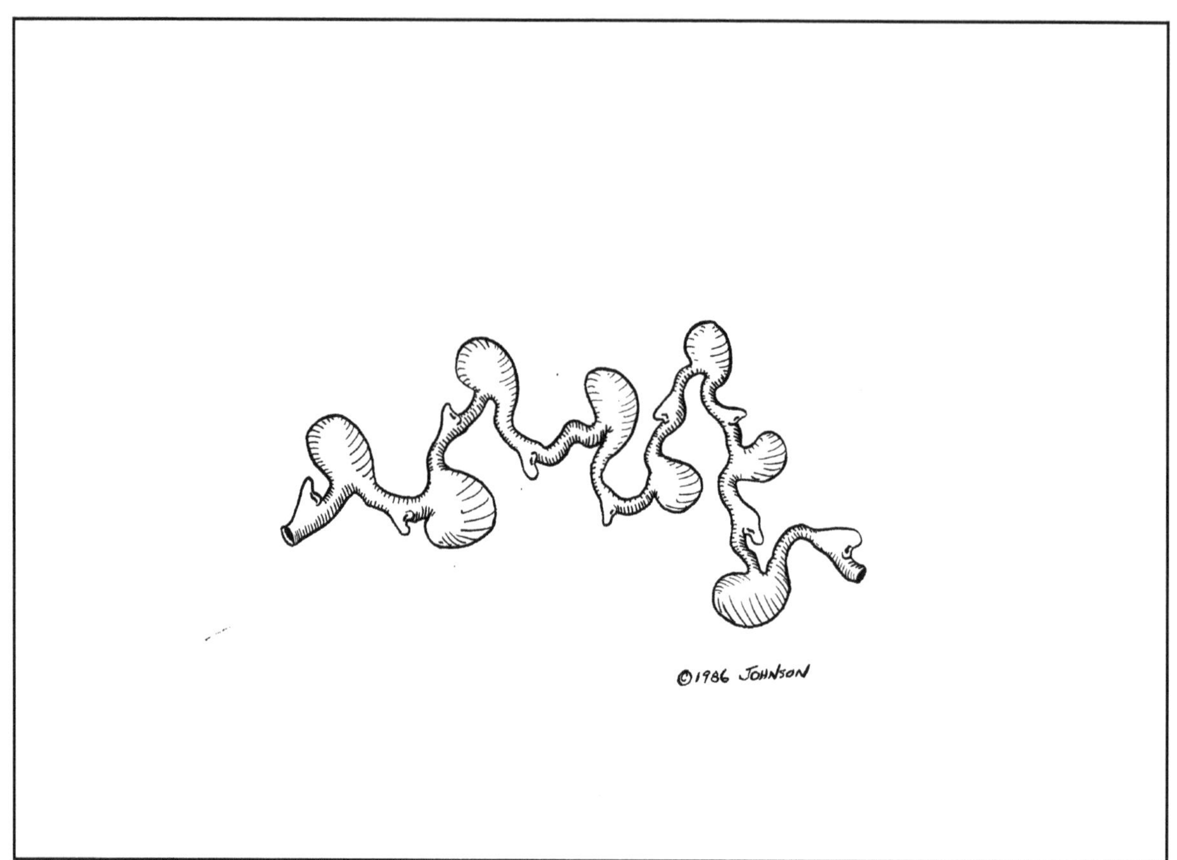

The Nasogastric Tube

A vital tool in gastrointestinal surgery, the nasogastric tube is used to keep the stomach decompressed.

Patients do not delight in it.

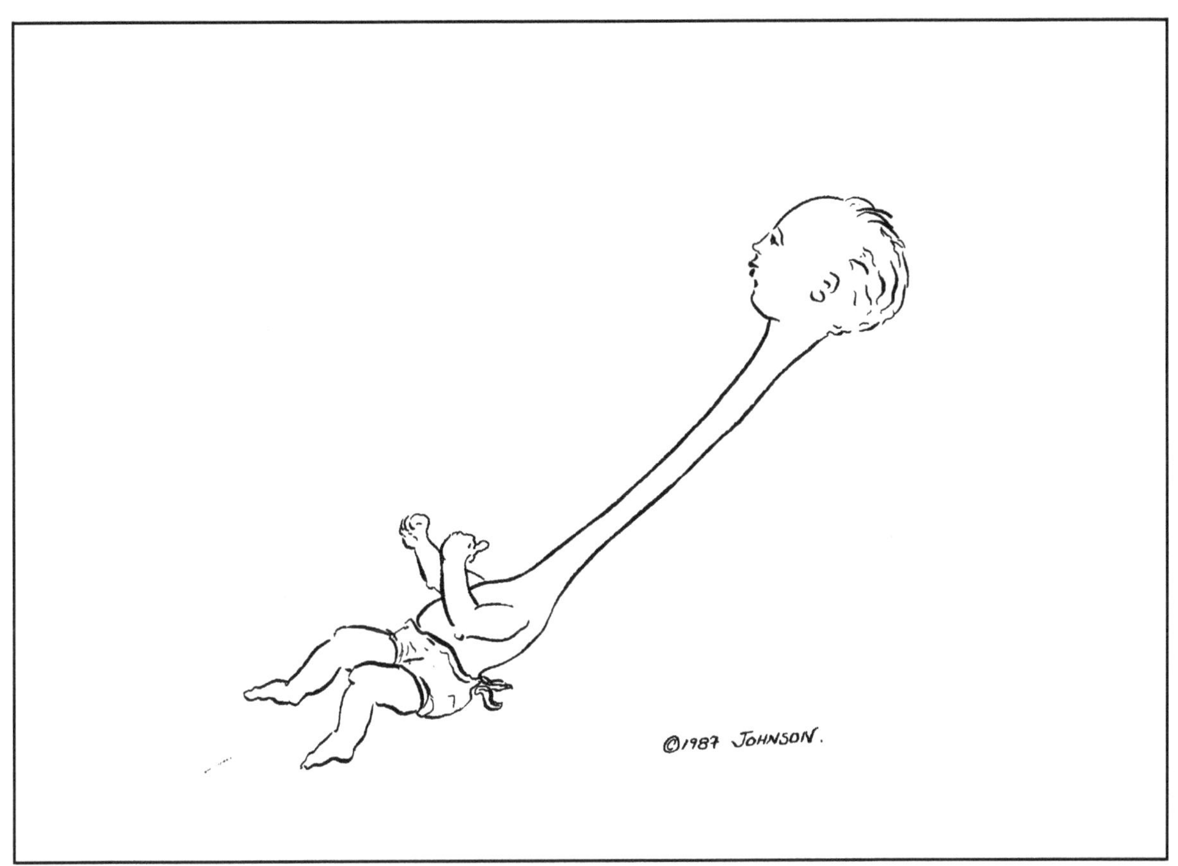

The NEC Baby

Necrotizing Enterocolitis (NEC) is a dangerous condition of small infants in which large sections of the intestines die. Only careful surgery and intensive care allow them to survive. They generally require hyperalimentation for long periods of time.

Normal Oral Flora

Our mouths are loaded with multiple forms of bacteria. When a culture of the mouth is taken and a general non-pathological pattern of bacteria is found, the lab report reads 'Normal Oral Flora.'

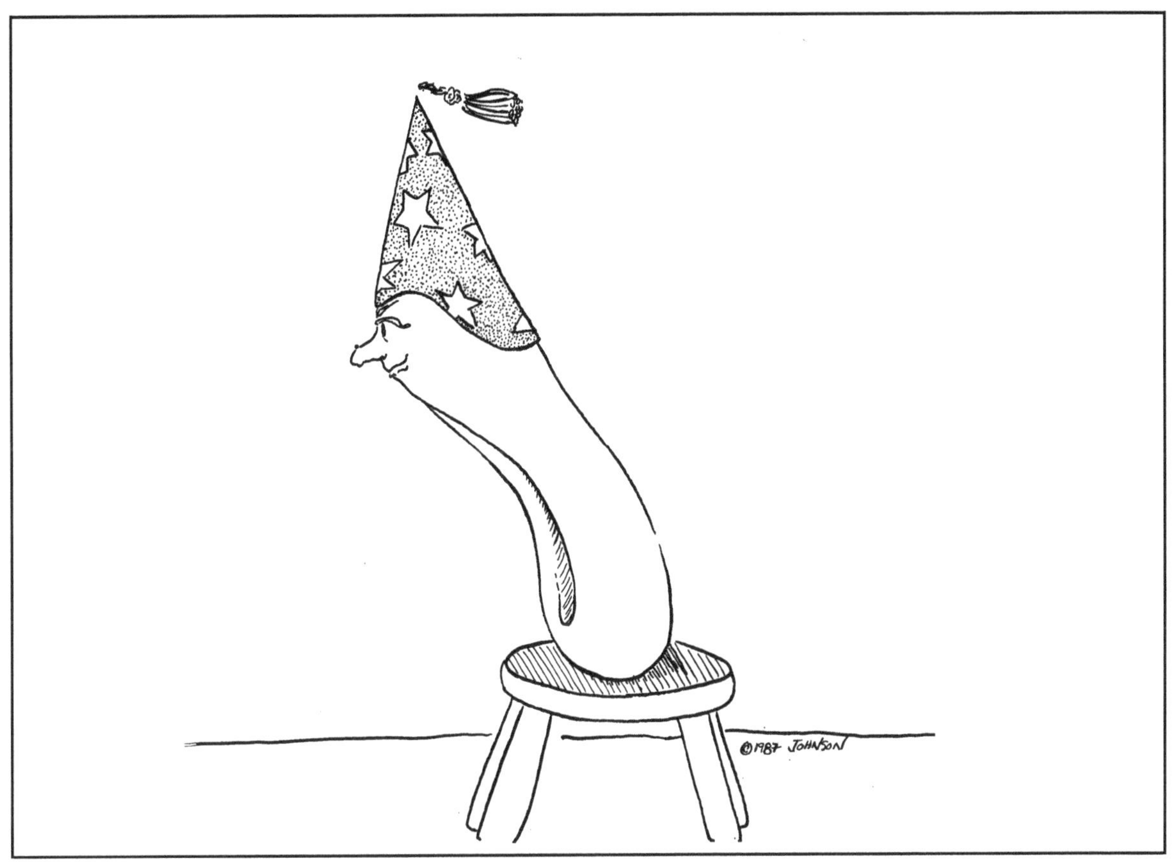

Occult Blood On The Stool

During a routine physical, a rectal exam is done and the physician touches the examining glove to a special card and adds a developer that can detect minute amounts of blood, invisible to the eye. When positive, occult blood is said to be on the stool. It is often a harbinger of colon cancer.

Passing A Swan

A Swan-Ganz catheter is inserted into the subclavian or jugular vein. Its tip is then floated through the right atrium and ventricle into the pulmonary artery. It is used to measure cardiac filling pressure using the wedge technique (described later).

Passing Gas

This is the exclusive domain of the anesthesiologist.

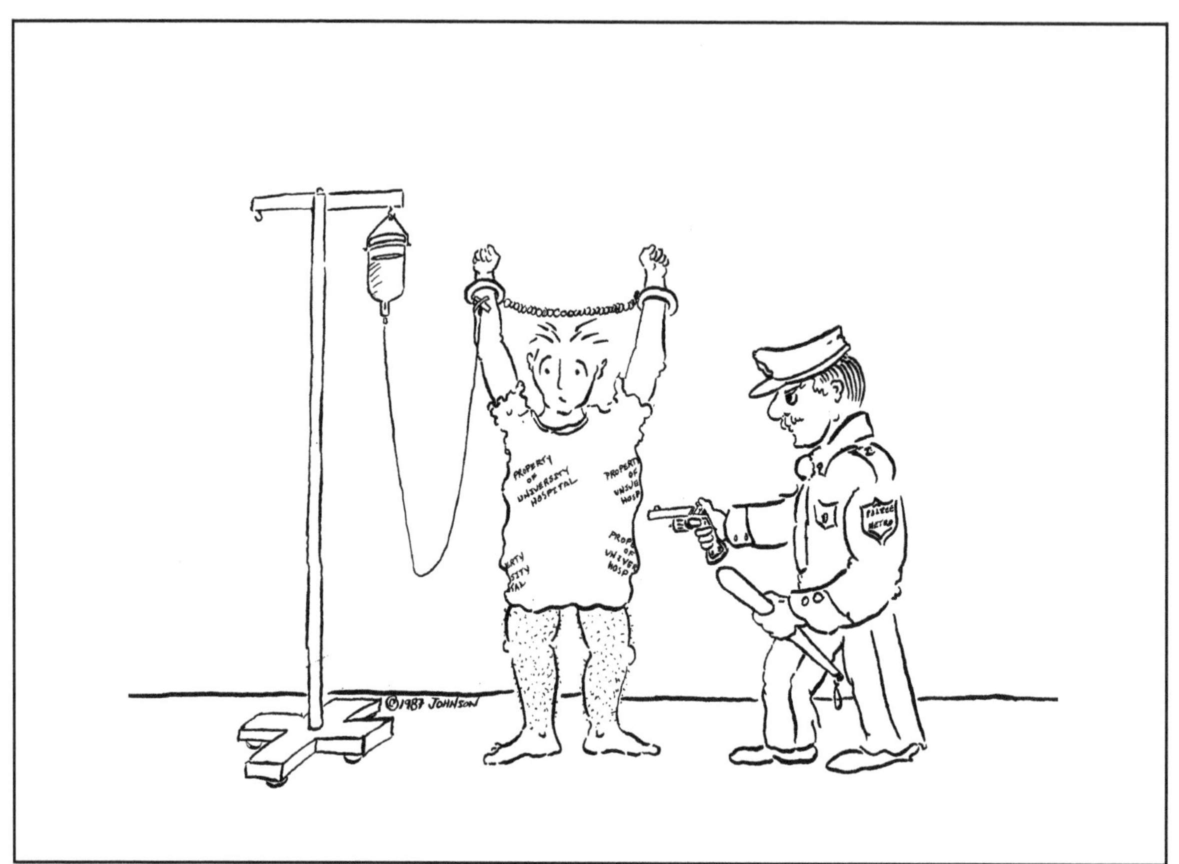

A Patient In Full Arrest

The patient who is neither breathing nor has a heartbeat is said to be in full cardiopulmonary arrest. Rapid CPR and/or defibrillation are often life saving.

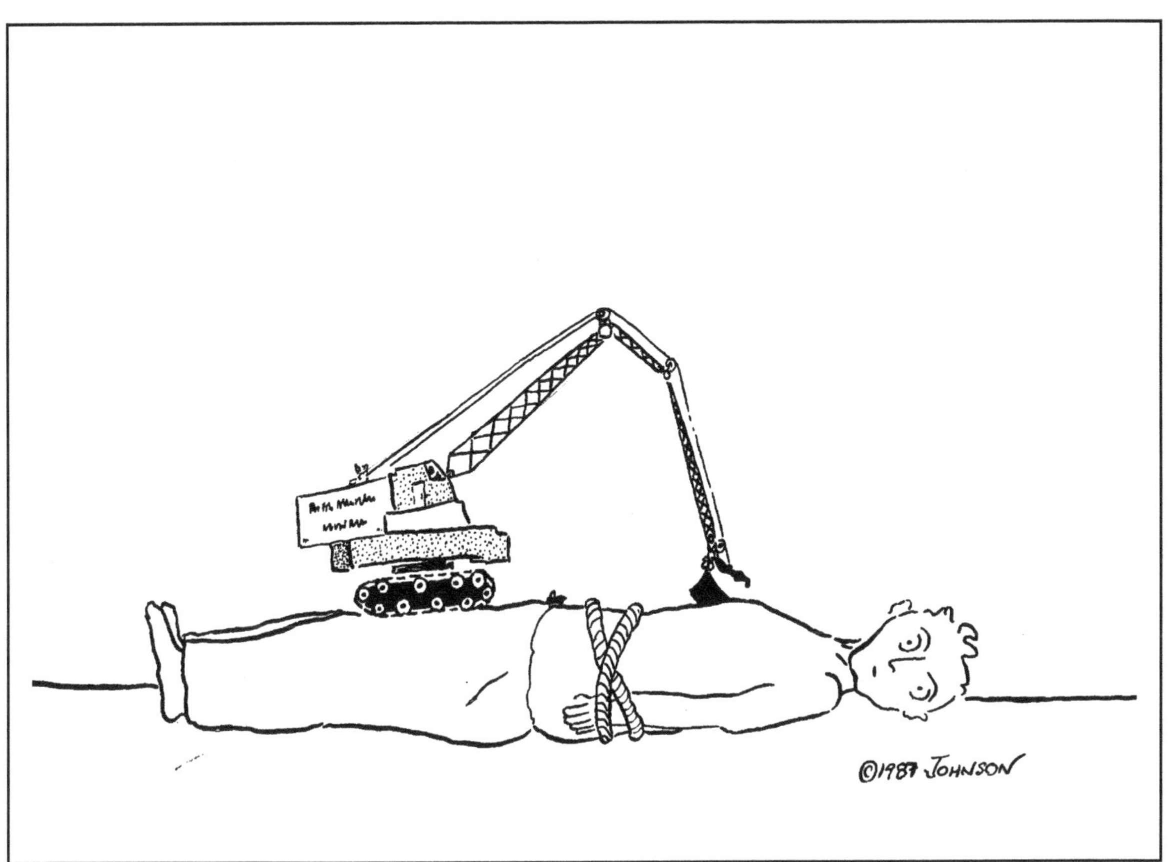

Pectus Excavatum

A birth defect of the chest wall, pectus excavatum is characterized by a dramatic inward relocation of the sternum, leaving a deep concavity. Because of the volume change in the chest, cardiac and pulmonary problems can be associated.

The Pediatric Surgeon

A surgeon who is focused on the operative care of children. The surgical conditions children face are often different from those of adults. Their operative and postoperative management require special skills - and a special attitude.

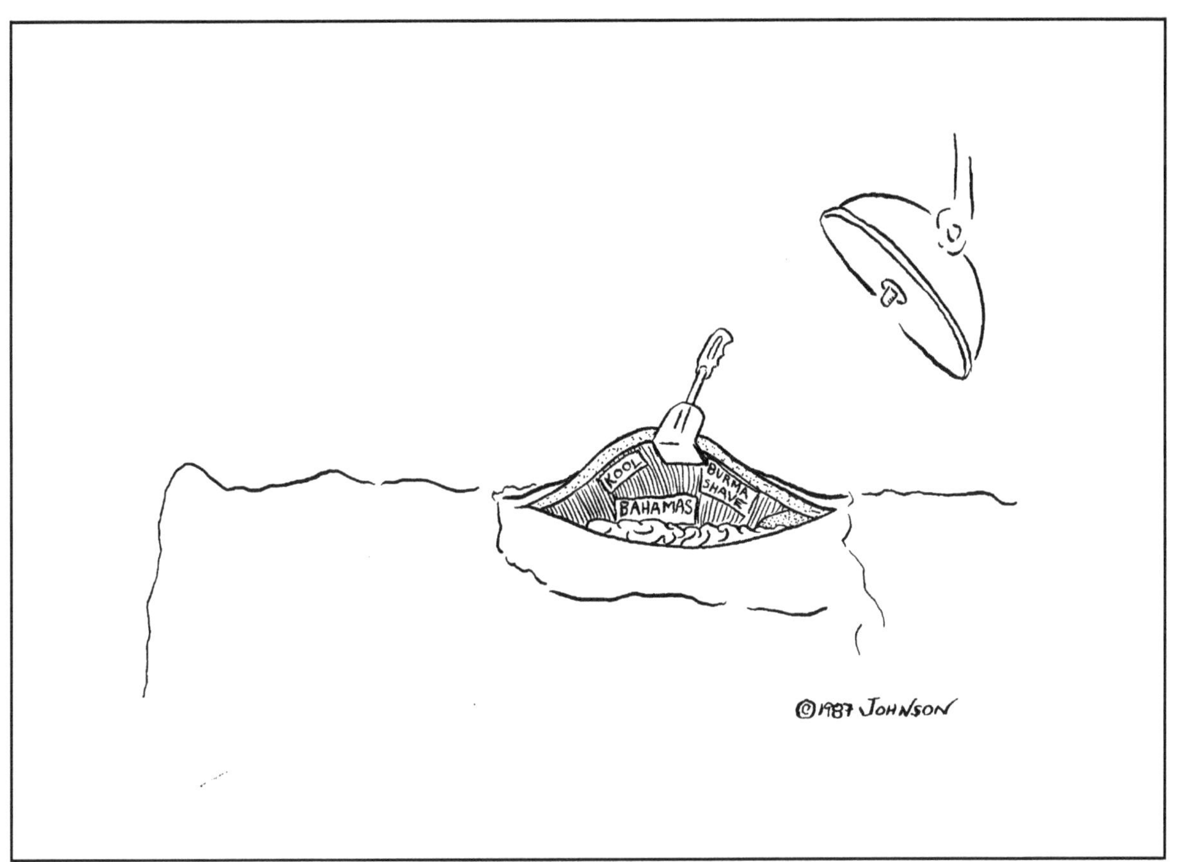

Peritoneal Signs

The inner lining of the abdomen, known as the peritoneum, is laden with sensory nerve endings. When inflammation is present, tenderness and reflex muscular rigidity are seen. These are known as 'peritoneal signs,' often present in appendicitis.

Picking Up Cases

A surgical resident has five years to acquire the experiences needed to become an independent, safe operative surgeon. The aggressive resident will seek out operations ('cases') to perform, usually by being extra cordial to the chief medical resident.

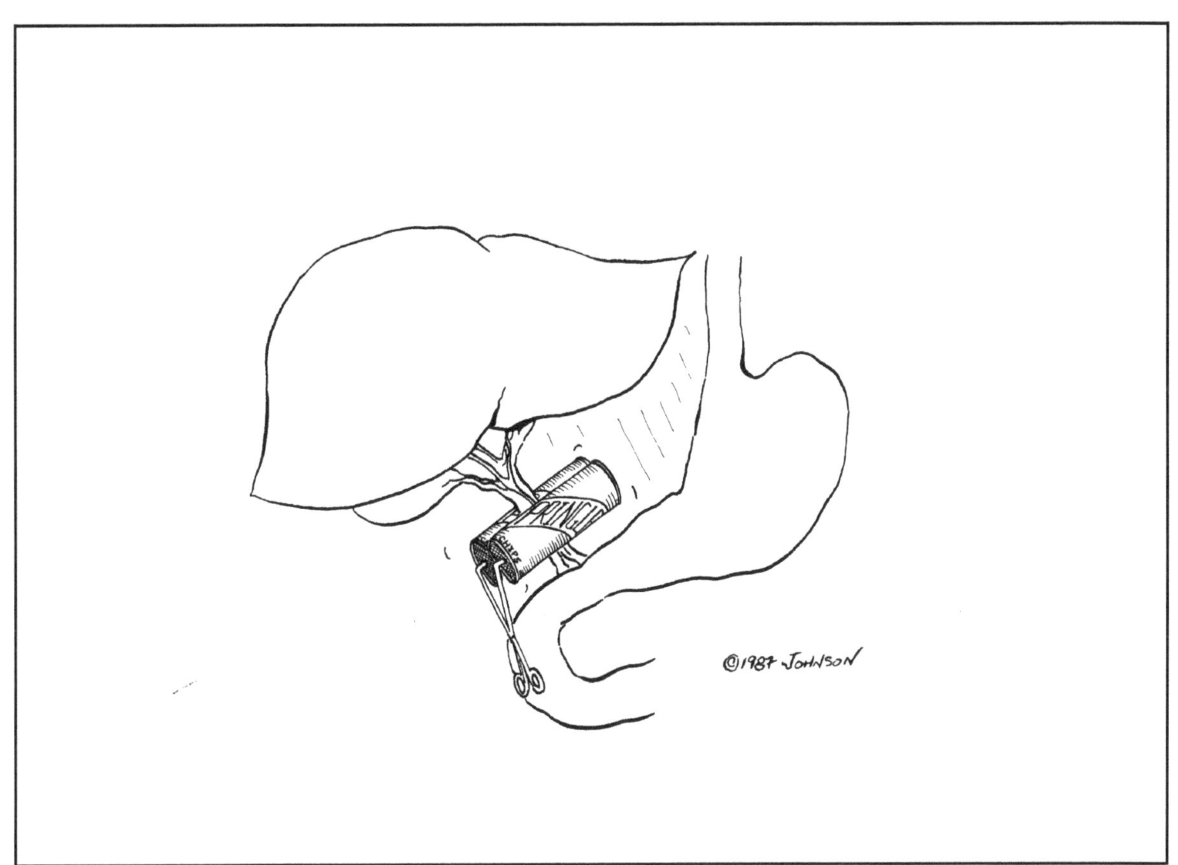

The Pringle Maneuver

The placement of an occluding clamp across the hepatoduodenal ligament, which contains the hepatic artery and portal vein is used to reduce blood flow to the liver after liver trauma or during a planned resection.

Projectile Diarrhea

Occasionally—and generally in children—diarrhea can be so explosive as to be expelled for long distances.

This is when a white coat comes in handy on rounds.

Pulmonary Toilet

When a patient has pneumonia or heavy bronchial secretions, he or she is positioned face down and the back is thumped repeatedly to release secretions and induce coughing to clear the lungs. The look on a new patient's face when told he will undergo 'Pulmonary Toilet' is truly a sight to see.

Putting Down A Nasogastric Tube

The bane of the intern—as well as the patient—is the nasogastric tube. Here, the intern is shown expressing his true feelings toward this otherwise highly useful device.

Placing The Patient On A Sliding Scale

Diabetics often require supplemental insulin to cover increases in their blood sugar after the stress of surgery. To enable the nurse to dose the insulin properly, graded amounts are prescribed as a function of the level of blood sugar — known as a sliding scale.

Recording The Stool

Patients with uncontrollable diarrhea (as in cholera, carcinoid syndrome and the like) often lose life-threatening amounts of liquid via the stool. An accurate accounting is essential to guide fluid replacement, generally by the IV route.

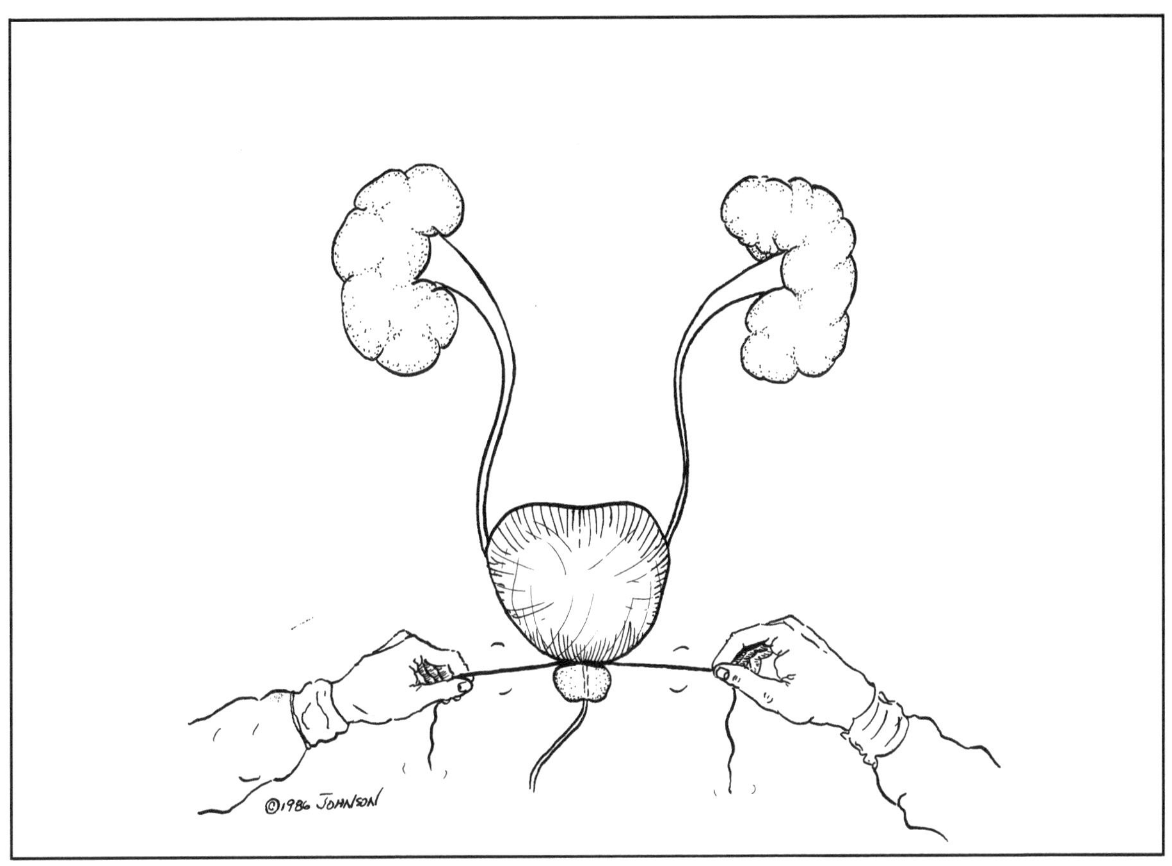

Retention Sutures

When the intestines are so swollen at the end of an operation that the abdomen cannot be easily closed, strong bridging sutures known as 'retention' sutures are used. However, another form of retention is urinary and it would definitely occur with the approach shown here.

Rhinoplasty

This is better known as a 'nose job.'

Securing An Airway

When a patient stops breathing, life depends upon gaining and securing access to the trachea for ventilation.

In this case, the surgeon has gone overboard.

Shooting A Scout

A preliminary x-ray, taken to determine that the patient is well positioned for the final x-ray, is termed a 'scout film.'

This image depicts one reason why surgeons do not perform x-rays.

Small Bowel Follow Through

In an upper GI series, the patient swallows radiodense barium and the esophagus and stomach are visualized. After a time, the barium enters the small intestine and can be followed until it reaches the colon. There is no rushing it.

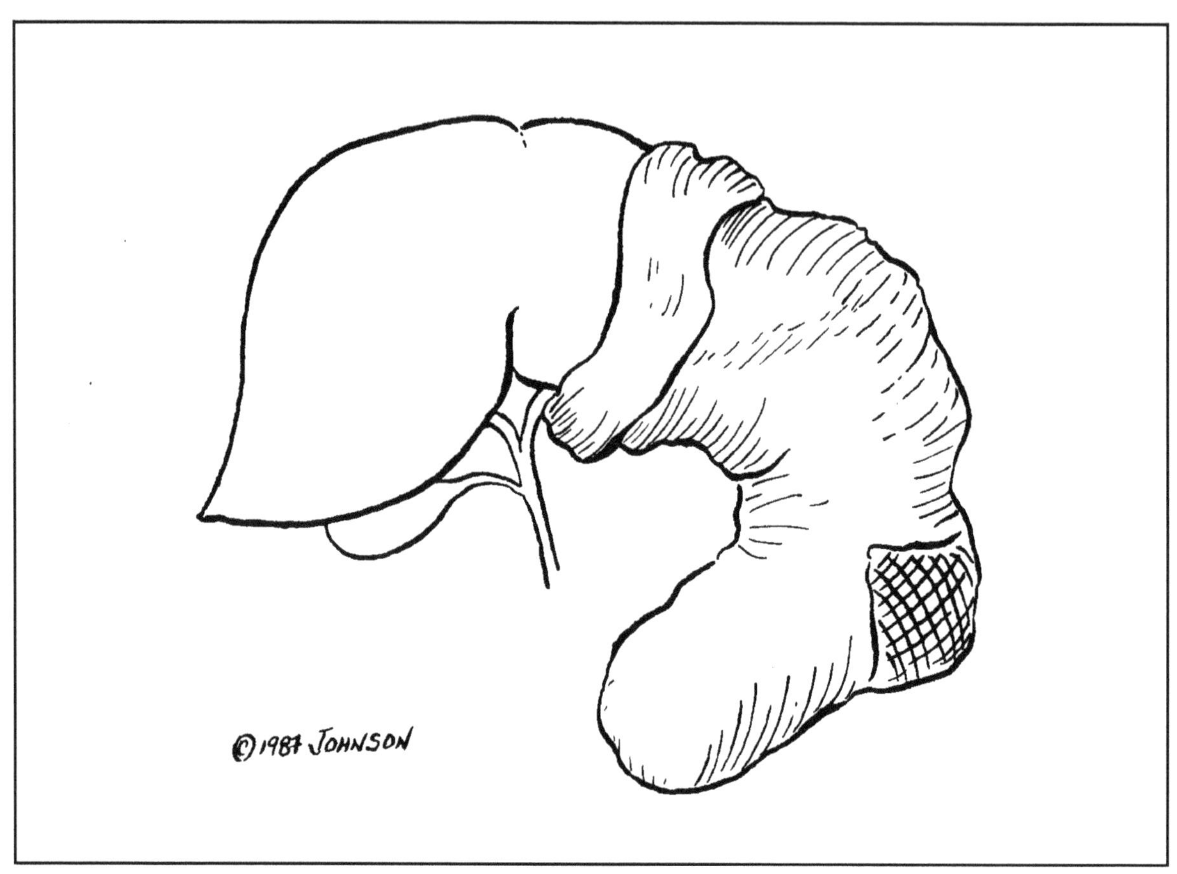

Socked In Liver

In reoperative surgery of the abdomen, scar adhesions are often extensive. When the liver is so encased in scar that dissection becomes tedious and dangerous, it is said to be 'socked in.'

Spilling Sugar In The Urine

Urine is normally free of sugar. When blood sugar is sufficiently elevated (in diabetes), sugar can often be detected in the urine. Medieval MDs often tasted urine to detect the condition.

This is frowned upon today.

Firing The Stapler

Incredibly ingenious stapling devices are now available to reattach transected structures such as blood vessels ducts and bowel. The positive click of the fired stapler is a very gratifying sound, as it has just saved an hour or more of suturing.

STAT Portable Chest

When the patient cannot be moved to the Radiology Dept. for a chest film, a portable machine is brought to the bedside and the x-ray is taken. When this is an emergency (i.e., 'stat'), the technician arrives somewhat more rapidly. ['Statim' (L) = immediately].

Stool Impaction

Every surgical intern fears this happening to the patient, as it may dictate manual disimpaction of the rectum. The latter is predictably associated with a thoughtful reappraisal of one's career choice.

Stool Softener

One of the most prescribed medications by interns, for previously described reasons.

The Sucking Chest Wound

The chest wall is an elastic enclosure that holds the lung open by creating an internal vacuum. If the chest wall is punctured, air rushes in, allowing the lung to collapse. The effect is the opposite of that which occurs after perforation of a pressurized aircraft's fuselage.

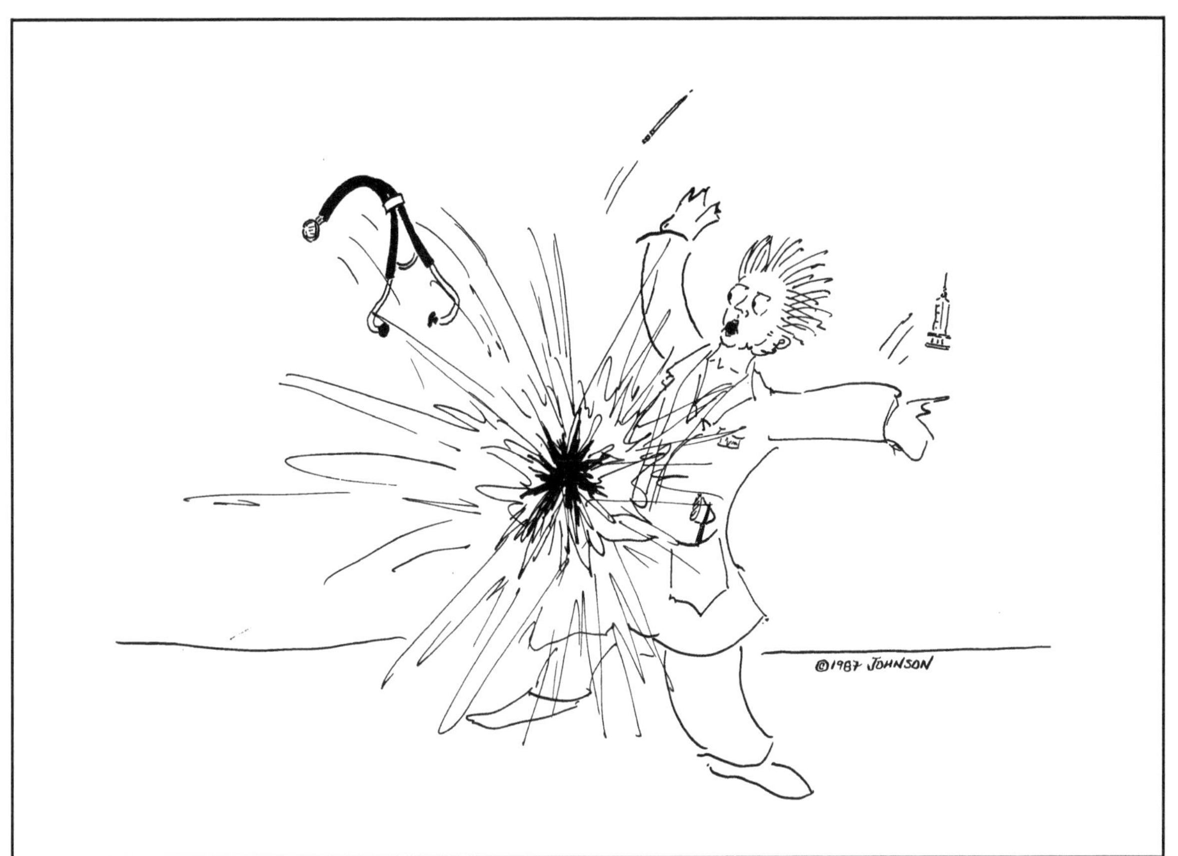

Suddenly, His Pager Went Off

Sometimes, this is exactly what it feels like.

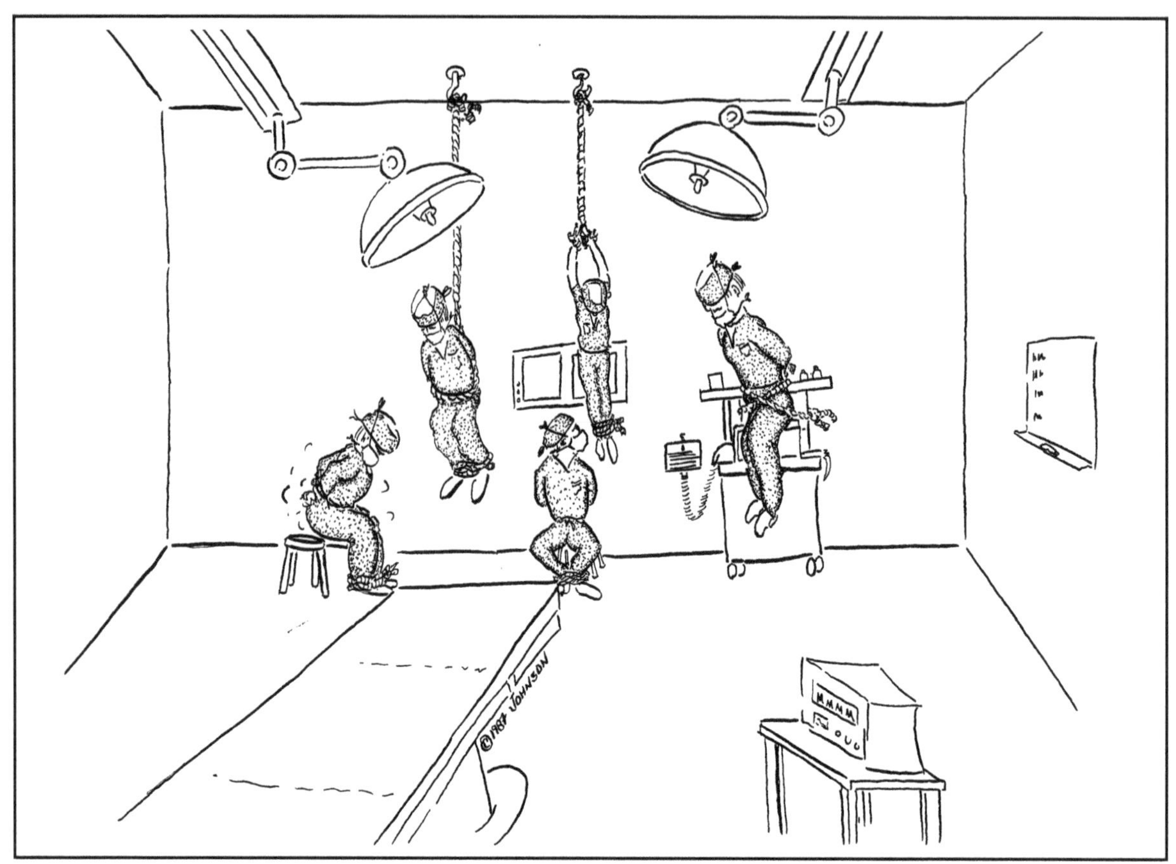

The Surgeons Are Tied Up In The OR

One of the great things about being a surgeon is that the office can always credibly state that you are 'tied up in the OR,' even if you are actually golfing.

Taking The Heart Back

In open heart surgery, the blood is anticoagulated heavily. If not fully reversed or if vessels are otherwise bleeding, the patient must be returned to the OR to remove clot and secure control.

Recently relieved families react poorly to this news.

Taking The Patient Down To Surgery

It is axiomatic that the Operating Room is always 'down' relative to where one is now.

3/4 Strength Breast Milk

The neonate with an intestinal disorder may not be able to digest full strength breast milk. It is watered down to match the child's digestive capabilities. This is prepared in a graduated cylinder after full strength breast milk is pumped, not as shown here.

Thyroid Storm

The extremely hypermetabolic condition that occurs when the thyroid gland secretes uncontrollable amounts of thyroid hormone.

The Tissues Were Friable

Patients taking steroids for long periods lose much tissue structure, making surgery far more difficult. Tissues that cannot hold sutures, for example, are considered 'friable.' This is a description that should be avoided, however, when in elevators or at the bedside.

The Trauma Service

Extremely well organized rapid response teams, Trauma Services at all major hospitals are often unsung heroes of patient care.

Upper GI Bleed

Bleeding from the esophagus, stomach or small intestine is termed an 'upper GI bleed,' as opposed to a 'lower GI bleed' from the colon. Diagnosis and treatment are challenging in each case.

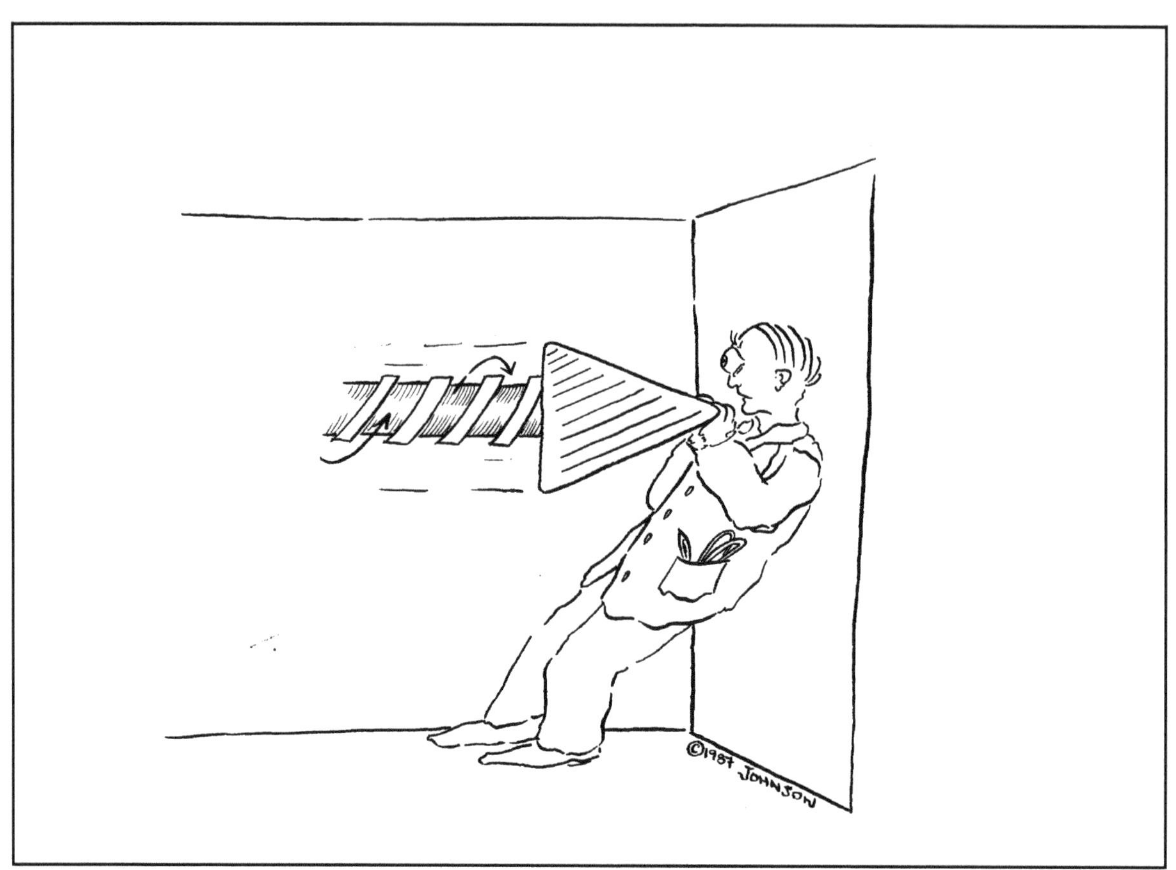

Wedge Pressure

The heart pumps as a function of the pressure of blood in its chambers. A way to measure this is to insert a pulmonary artery catheter (i.e., Swan-Ganz catheter), expand a small balloon and allow it to wedge in a small lung artery. The tip then 'sees' the static pressure that fills the heart.

The Well Baby

A child who, by definition, does not yet require surgery.

Appendix

(and no, I do not mean the real one).

The final section of this book is devoted to the training of a surgeon throughout his or her residency.

It was this process that begat the twisted notions that led to these cartoons.

Training A Surgeon

Surgical training is a marathon, not a sprint. It is an exceptionally demanding process that traverses multiple levels of personal development.

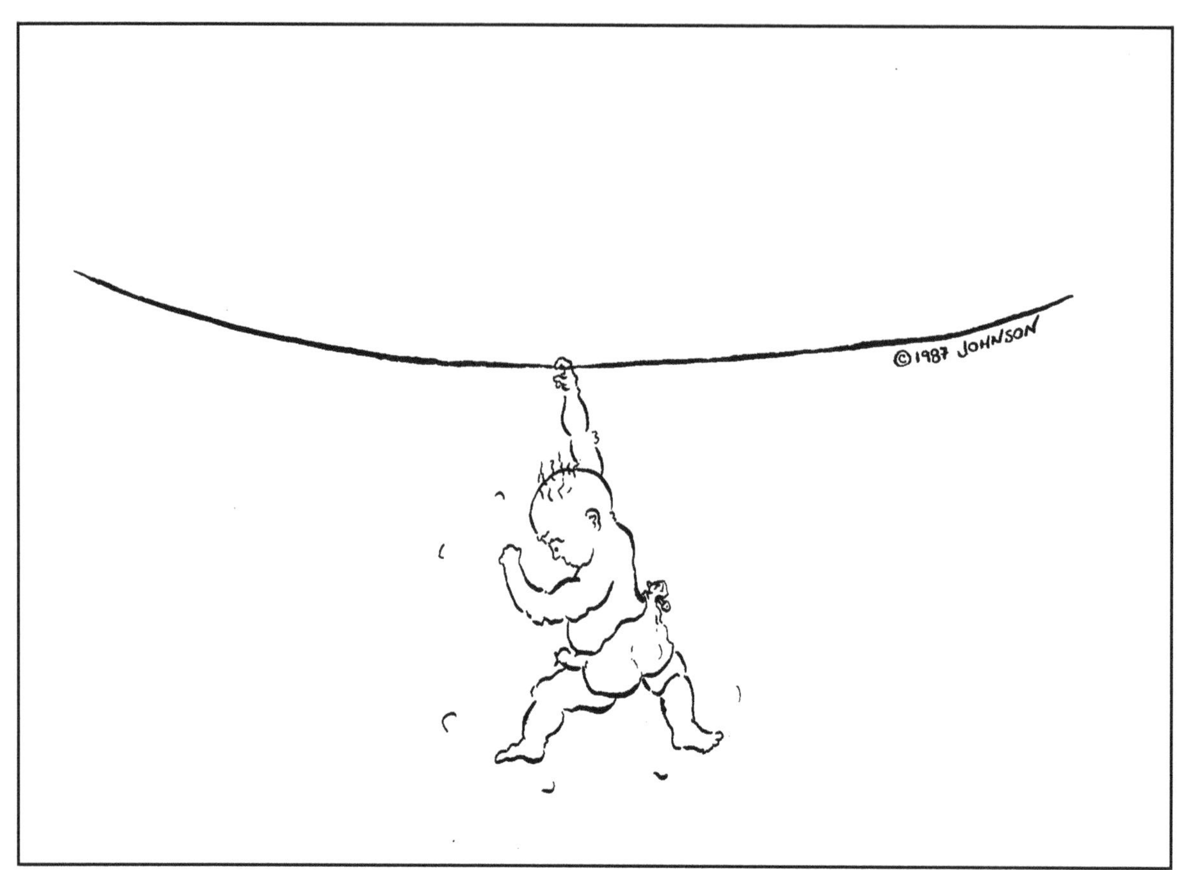

The Intern

The surgical intern rapidly finds that he or she has much to learn. Hanging on is a victory in itself. The intern is usually found in the wards, managing postoperative patient care.

Occasionally, the Chief Resident will call the intern to the operating room to place a stitch or to write nursing orders.

The Second Year Resident

While primarily responsible for supervising interns on the ward, the second year resident may actually perform simpler operations under the watchful eye of the skeptical Chief Resident.

The Resident In The Research Years

Often, surgical training programs allow their residents to take one to two years off to perform research projects. This is where the surgeon learns that six months in the lab can save an afternoon in the library.

The Third Year Resident

Where surgical hubris begins, the third year resident is less supervised, runs clinics, manages ward teams and performs operations of moderate difficulty routinely. A false sense of capability and security occur at this stage.

The gods do this on purpose.

The Fourth Year Resident

The beginnings of surgical maturity occur here. Exposed to more complex consulting and leadership tasks, the 4th year resident has felt the sting of bad decisions and the wrath of the M&M conference. He or she becomes more careful. Aging occurs.

The Chief Resident

Seen as all-knowing by junior residents, the Chief Resident learns more in this year than in any other. Now fully responsible for all patient care decisions, it is in this year that the career tenor of the surgeon is established.

A glimmer of humility begins.

Board Certified

The affirmation of one's successful training by peers, Board Certification is a peak moment in the surgeon's life.

From then on, he or she is a real member of the surgical fraternity.

I'm On My Beeper

The excellent surgeon is continuously available—on beeper, cell phone, or PDA. Such commitment is rare in society and does not do much for the development of involved hobbies, like cartooning, for instance.

But it is appreciated.

About The Author

Peter C. Johnson, MD, a former academic surgeon, is the President and CEO of Scintellix, LLC, a biomedical technology consultancy.

He is also is Principal, Headway LifeScience Resources, a life sciences executive search firm.

He lives with his family in Raleigh, NC.

www.ingramcontent.com/pod-product-compliance
Ingram Content Group UK Ltd.
Pitfield, Milton Keynes, MK11 3LW, UK
UKHW051128260726
13967UKWH00010B/2930

9 781435 717206